J. WESTON
WALCH
PUBLISHER
Portland, Maine

Content Reading Strategies

Mathematics

Josh Brackett

User's Guide
to
Walch Reproducible Books

Purchasers of this book are granted the right to reproduce all pages where this symbol appears.

This permission is limited to a single teacher, for classroom use only.

Any questions regarding this policy or requests to purchase further reproduction rights should be addressed to:

> Permissions Editor
> J. Weston Walch, Publisher
> 321 Valley Street • P.O. Box 658
> Portland, Maine 04104-0658

1 2 3 4 5 6 7 8 9 10

ISBN 0-8251-4335-7

Copyright © 2002
J. Weston Walch, Publisher
P.O. Box 658 • Portland, Maine 04104-0658
www.walch.com

Printed in the United States of America

Contents

Introduction: To the Teacher .... *v*
Introduction: To the Student .... *vii*

Part 1: Building Vocabulary 1
    Lesson 1: Using Context Clues 2
    Lesson 2: Prefixes and Suffixes 8
    Lesson 3: Word Roots 12

Part 2: Prereading 15
    Lesson 4: Previewing 16
    Lesson 5: Predicting 21
    Lesson 6: Prior Knowledge 22
    Lesson 7: Purpose 23

Part 3: Reading Strategies 25
    Lesson 8: Introduction to Reading Strategies 26
    Lesson 9: KWL 27
    Lesson 10: SQ3R 32
    Lesson 11: Semantic Web 38
    Lesson 12: Outline 43
    Lesson 13: Structured Notes 48

Part 4: Postreading 57
    Lesson 14: Summarizing and Paraphrasing 58

Part 5: Reading in Mathematics 63
    Lesson 15: Common Patterns and Features of Mathematics Writing .. 64
    Lesson 16: Main Idea and Details 65
    Lesson 17: Visual Texts 67
    Lesson 18: Concept and Definition 70
    Lesson 19: Concept on Concept 71
    Lesson 20: Steps in a Process 73
    Lesson 21: Review 75

Part 6: Practice Readings 77
    Practice Reading A: Supply and Demand 78
    Practice Reading B: The Three Suspects 81
    Practice Reading C: Sampling Red Blood Cells 83
    Practice Reading D: Archimedes 86

Blank Graphic Organizers 89
    4-P Chart 90
    KWL Chart 91
    SQ3R Chart 92
    Semantic Web 93
    Outline 94
    Structured Notes 95

Teacher's Guide and Answer Key 97

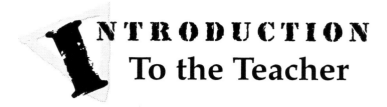

Introduction
To the Teacher

Content-Area Reading Strategies teaches students how to read to learn. In the early grades, students learn to read and write narratives—stories. They are used to dealing with texts that have a beginning, a middle, and an end. They expect to encounter rising action that leads to a climax and then to a resolution.

This pattern of organization is often not followed in informational texts, which begin to make up a large portion of classroom-related reading about grade four. Without instruction in how to read these kinds of nonnarrative texts, even "good" readers can stumble. Some research shows that the so-called "fourth-grade reading slump" may be attributable in part to the unsupported transition from narrative to informational texts.

That's where the *Content-Area Reading Strategies* series comes in. Each book in the series focuses on a different content area, and gives students concrete tools to read information texts efficiently, to comprehend what they read, and to retain the information they have learned.

Organization is an important part of comprehending and retaining knowledge. The graphic organizers in *Content-Area Reading Strategies* help students connect new information to their existing schemata, increasing their ability to recall and take ownership of what they read. The reading strategies give students a way to "see" what they read—a great asset to visual learners.

The reading-writing connection is a strong one. The reading strategies in this book all require students to record information in writing, strengthening readers' ability to retain and access newly acquired knowledge.

Classroom Management

Content-Area Reading Strategies is easy to use. Simply photocopy each lesson and distribute it. Each lesson focuses on a single strategy and includes models showing the strategy in action. At the back of this book, there are blank copies of each graphic organizer, so you can copy them as often as needed. Quiz questions assess how well students understood what they read.

The Practice Readings provide longer readings and questions. For these, you may want to let students choose which strategy to use, or you may assign a particular strategy. Either way, have copies of the appropriate graphic organizers available.

Eventually, students will no longer need printed graphic organizers; they will make their own to suit their learning style and the particular text they are reading. They will have integrated the reading strategies as part of the learning process in all content areas.

INTRODUCTION
To the Student

Welcome to *Content-Area Reading Strategies*! This book will give you the tools you need to read, understand, and retain the texts you read for school. These strategies will save you time by helping you to extract information efficiently from your reading, to organize that information so that it is easy to understand, and to remember the information through writing it.

You may have heard about the writing process—the steps you follow to write a paper, letter, story, or anything else. But did you know there is also a reading process?

The reading process consists of three major steps: prereading (before reading), during reading (the reading itself), and postreading (after reading). This book gives you specific strategies to use to accomplish each step. These strategies are given form in graphic organizers, which ask you to think and write about your reading. Breaking up a big task, such as reading a whole mathematics chapter, into smaller steps makes the job easier to tackle. These graphic organizers make studying for tests less stressful, too—all the information is already written down in a condensed form, in your own words. No more last-minute cramming!

Everyone reads and learns differently, and you will probably find that some strategies are more helpful to you than others. Give them all a try, and find out what works for you. Customize the strategies to your way of reading and learning. Eventually, you will not need the printed graphic organizers at all. You'll follow the reading process steps automatically and organize the information in the way most meaningful to you.

PART I
Building Vocabulary

Lesson 1
Using Context Clues

There are about 750,000 words in the English language. No one knows them all. As long as you read, you will come across words you don't know.

If you can learn how to decipher the meaning of unfamiliar words, several good things will happen. You'll not only understand and remember what you're reading, you'll also learn to recognize and understand new words. And you'll acquire a larger vocabulary of words you can use effectively.

Context Did you know the word *decipher* before you started reading this lesson? (It appears in the second paragraph of this page.) Even if you didn't, you probably understood the paragraphs anyway. How? You knew you were reading about reading strategies. You knew it meant something you do to unfamiliar words. You probably inferred from the context that *decipher* means something like "decode" or "figure out." And you were correct.

The *context* of a word is all the words that come before and after the word. Graphic elements like pictures, charts, and tables are part of the context of a word, too. To *infer* means to reason from evidence to a conclusion. To infer the meaning of a word from its context is to look at the evidence of the words and graphic elements around it and draw a reasonable conclusion about what the word means.

Context Clues Some types of context clues that can help you infer meaning are
- restatement
- explanantion
- description
- definition
- antonyms and synonyms
- examples

Your own experience gives you context clues, too. Things you have read, activities you have participated in, movies you have watched—just about anything you have stored in your memory—give you background to draw on when you meet a new word.

Using Context Clues (continued)

Context Clues in Action

Look at the passage below. You will see that some words are underlined. Let's suppose that the underlined words are completely unknown to you. Read the passage, and then read the paragraphs that follow it. These paragraphs are an example of the thought process you might go through to decipher the unfamiliar words.

Pendulums

Galileo Galilei, who lived in Italy from 1564 to 1642, used mathematics to make important contributions to physics and astronomy. One of the <u>phenomena</u> he studied was the behavior of <u>pendulums</u>. He noticed that if you tie a weight to the end of a string, hang it from a fixed point, and start it swinging, it will swing in a definite rhythm. Each swing <u>cycle</u> of the pendulum—from one side to the other and back—always takes the same amount of time.

Galileo also noticed that the time each swing cycle takes has nothing to do with the weight of the pendulum or where it starts. Instead, the <u>period</u> of a pendulum swing is a <u>function</u> of its length. If you change the length of the string, the period of the swing changes accordingly. The following table <u>expresses</u>, or shows, that function. (In Galileo's time, they used different units of length from the ones we use today.)

As you can see from the table below, in order to slow the period of the pendulum swing from 1 to 2 seconds, you have to more than double the length of the string; you have to quadruple it. In fact, you will note that each number in the length column is the square of the <u>corresponding</u> number in the period column. If you lengthened the string to 25 units, the period would be 5 seconds. The mathematical way to express the relationship between the length (in Galileo's units) of a pendulum, l, and its swing period (in seconds), p, is $p^2 = l$.

Length of Pendulum (in units)	Period of Swing (in seconds)
1	1
4	2
9	3
16	4

Using Context Clues *(continued)*

phenomena. Phenomena are what Galileo studied. He was into physics and astronomy. Must mean things you study in physics or astronomy. The behavior of pendulums is an **example.**

pendulums. Explained in the next sentence with a **description.** "He noticed that if you tie a weight . . ."

cycle. A **definition** is given between the dashes: "from one side to the other and back."

period. Seems to refer to "the time each swing cycle takes." Does *period* in other contexts—school, hockey, paleontology—mean a stretch of time? Yes. I used my own **experience** to help with this definition!

function. The sentence with *function* in it, "Instead, the period of a pendulum swing . . ." refers to the previous sentence. The words "has nothing to do with," taken with the word "instead," gives me the idea that function must mean "has to do with"—an **antonym/synonym** type of clue. Also, the table shows what the relationship between length and period is—what they have to do with each other.

expresses. This word is followed by a **restatement:** "shows."

corresponding. The sentence refers to the table, which shows $1=1^2$, $4=2^2$, $9=3^2$. Must mean the number that goes with the first number. This seems to be a kind of **example.**

Application The article on the next page has some words in it that you probably don't know. Use the context clues strategies to decipher them. If you wish to follow the directions and actually draw the design, go ahead, but the important thing is to understand the vocabulary of the article. After reading the article, answer the questions that follow it.

Using Context Clues (continued)

How to Draw an Ancient Islamic Design

Many of the most beautiful mosques of the Muslim world were built long before the decimal system of numerals was developed or units of measure were standardized. Yet these houses of worship are famous for their intricate tessellated designs.

In the Islamic faith, representations of humans or animals are forbidden in religious centers. Also, it is considered desirable for all the parts of a building—and, if possible, all the buildings of a community—to harmonize with each other. These principles led to the use of geometric patterns based on the repeated use of a standard circle.

By following the steps below, you can learn to use a standard circle to create a typical Islamic design pattern. Use a pencil rather than a pen. Draw lightly. Later, you will need to erase some of the curves you draw.

1. Draw a circle whose radius is about half an inch.
2. Draw another circle with equal radius whose center is on the first circle. Label one of the points where the two circles intersect A.
3. Draw a third circle with equal radius whose center is A.
4. Draw four more circles with equal radii through A whose centers are points where two other circles intersect.
5. The circle whose center is A now has six points on it where the other circles intersect it. Erase all of the other circles but leave the intersection points.
6. Draw line segments connecting nonadjacent intersection points to form two overlapping triangles.
7. Starting at one of the intersection points, erase the line segment that connects it to the triangle of which it is not a part.
8. Starting at each of the remaining intersection points, erase the corresponding line segment.
9. Erase the circle.
10. Draw the figure again so that the endpoints of two segments coincide. Repeat step 10 as many times as you wish.

This algorithm produces a tessellation of hexagons and parallelograms that is characteristic of traditional Islamic design.

Using Context Clues *(continued)*

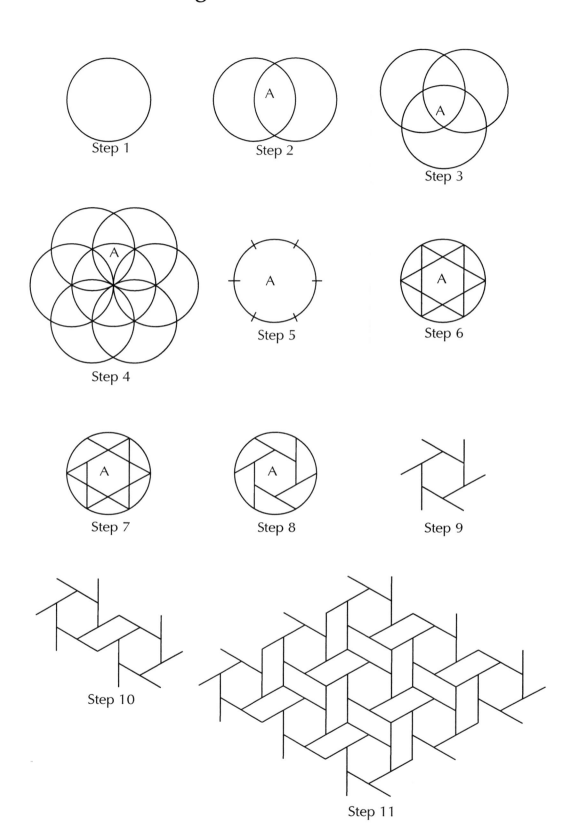

Using Context Clues (continued)

Circle the letter of the answer that best completes each sentence. Then explain briefly why you think your choice is correct.

1. An *intersection* is
    (a) a circle that overlaps another circle
    (b) a line or curve that contains two points
    (c) a point where two lines or curves cross
    (d) the inside of something that has been cut
    _____

2. *Standard* means
    (a) $\frac{1}{2}$-inch radius
    (b) established and agreed upon
    (c) Islamic
    (d) round
    _____

3. *Nonadjacent* means
    (a) connecting
    (b) next to each other
    (c) not connecting
    (d) not next to each other
    _____

4. An *algorithm* is a
    (a) step-by-step procedure
    (b) geometric design
    (c) polygon
    (d) house of worship
    _____

5. *Tessellation* means
    (a) covering a surface completely with overlapping
    (b) covering a surface completely without overlapping
    (c) covering a surface incompletely with overlapping
    (d) covering a surface incompletely without overlapping
    _____

Lesson 2
Prefixes and Suffixes

In addition to using context clues, you can use other strategies to figure out the meaning of new words. One of these strategies is to break down the word into its basic parts. Two of these parts are **prefixes** and **suffixes.**

What Are Prefixes?

Prefixes are word parts added to the beginning of a word that change the meaning of that word. For example, you are probably familiar with the verb *view*. If you add the prefix *re-*, which means "again," to *view*, you get *review*, which means "to view again." If you add the prefix *pre-*, which means "before," you get *preview*. *Preview* means "to look at before." You preview a movie and decide if it's worth seeing. You preview a novel by looking at the back cover and flipping through it to see if it is interesting enough to spend time on. And you preview informational texts to get an idea of what the assignment is about.

What Are Suffixes?

Suffixes are word parts added to the end of a word that change the meaning of the word. If you add the suffix *-er*, which means "one who," to the verb *read*, you get *reader*, "one who reads." Similarly, if you add *-ish*, which means "having the qualities of" to *child*, you get *childish*, which describes someone who behaves like a child.

Common Prefixes

Here's a list of common prefixes with their meanings.

Prefix	Meaning
anti-	against
co-, con-	with, together
e-, ex-	out of, away, from
i-, in-	in, into
in-, im-	not
inter-	between
intro-, intra-	in, into, inside
mis-	bad, wrong
multi-	many
non-	not

Prefix	Meaning
o-, ob-, oc-, of-, op-	toward, against
post-	after
pre-	before
pro-	for, forward
re-	back, again
sub-	below
super-	above
tele-	distant, far
un-	not

Prefixes and Suffixes (continued)

Common Math Prefixes

Here are some prefixes whose meanings are numerical.

Prefix	Meaning
femto-	quadrillionth
pico-	trillionth
nano-	billionth
micro-	millionth
milli-	thousandth
semi-, hemi-	half
mono-	one
bi-	two
tri-	three
quad-, tetra-	four
quint-, penta-	five

Prefix	Meaning
hex-	six
sept-	seven
oct-	eight
non-, nov-	nine
dec-	ten
cent-	hundred
mil-, kilo-	thousand
mega-	million
giga-	billion
tera-	trillion

Common Suffixes

Suffix	Meaning
Adjective Suffixes	
-able, -ible	capable of, worthy of
-ac, -al, -an, -en, -ent, -ic, -ine, -ish, -ive, -ous, -ious	of, like, relating to, being
-ful	having, resembling, able to
-less	without, lacking
-ly	having the quality of
Adverb Suffix	
-ly	in the manner of
Noun Suffixes	
-an, -ian	of, from
-ant, -ent, -ar, -ary, -er, -ess, -ist, -or	one who, that which
-ance, -ence	state or condition, act, quality
-ancy, -ency	state of, condition
-ary, -arium, -ory, -orium	place or device containing or associated with
-hood, -ness	state, quality of, condition
-ion, -tion, -ity	state, quality, act of
-ism	quality, doctrine, theory, system
-ment	act, state
-ship	state, quality, condition
Verb Suffixes	
-ate, -ify	make, act, cause to become
-en	make, become, cause to have
-ise, -ize	cause to be, make, act

Prefixes and Suffixes (continued)

Prefixes and Suffixes in Action

If you learn to look for prefixes and suffixes, you can break down a word into the parts that make it up, then use the context to figure out what the main part of the word means. Taken together, this information can help you understand the new word.

Read the paragraph below. Then read the thought process that a reader might use to figure out the meaning of the underlined words.

> By volume and weight, more than half of the structure of living creatures on Earth is made up of water. Our human bodies are more than 60 percent water! Our planet is inhabited by <u>hydraulically</u> designed <u>organisms</u>.

> One of the underlined words is *organisms*. I already know what it means, but when I break it down, I get a more precise meaning. *-Ism* means "system," and added to *organ*, I guess it means "a system of organs." I never thought of it like that before.
>
> *Hydraulically* is a long word! The first part of the word reminds me of *hydrant*, and the paragraph talks about water, so I think that the word has something to do with water. *-Ic* and *-al* mean "of" or "relating to," and *-ly* is an adverb suffix meaning "in the manner of." *Hydraulically designed organisms* must mean organ systems that have been designed in a manner relating to water.

Application

Read the following paragraph. Use prefixes, suffixes, and context clues to figure out the meanings of unfamiliar words.

> All structures are made stable through a balanced interaction between tensive and compressive forces. In nature, crystalline, cohesive bonds resist tension. Liquids resist compression. Being flexible, liquids spread their loads evenly to all the surfaces of their tension-resistant containers. As long as the container is strong enough, the system will hold its shape, because the contained liquids, which entirely fill the container, are noncompressible.

Prefixes and Suffixes (continued)

Circle the letter of the answer that best completes each sentence. Then explain briefly why you think your answer is correct.

1. *Interaction* is action
   (a) between
   (b) into
   (c) away from
   (d) toward
   Explanation: _____

2. *Tensive forces* are forces that
   (a) pull
   (b) push
   (c) squeeze
   (d) explode
   Explanation: _____

3. *Compressive forces* are forces that
   (a) pull
   (b) push
   (c) squeeze
   (d) explode
   Explanation: _____

4. *Crystalline bonds* are
   (a) cold
   (b) hard
   (c) hot
   (d) soft
   Explanation: _____

5. *Cohesive* means
   (a) cooling off
   (b) falling apart
   (c) heating up
   (d) sticking together
   Explanation: _____

Lesson 3
Word Roots

In Lesson 2, you learned to decipher unfamiliar words by breaking them down into parts. You have a list of prefixes and suffixes that are commonly used in English and some that appear often in mathematical contexts.

In this lesson, you'll learn some roots. These are the word parts that carry the main meaning of the word. The roots in the list below come up often in the reading you will do for school or work, or are used in mathematics.

Root	Meaning
angle	corner, turn
astro, aster	star, outer space
audi	hear, sound
bio	life
chron	time
cir, circul	circle
congru	come together
cur	run
dia	across, through
dict	speak
distrib	divide, scatter
duc	lead, bring
equ	equal
fac, fic, fec, feas	do, make
flect, flex	bend, contract
form	shape
fract, frag	break
geo	earth
gon	angle
gram, graph	write
gress, grad	step
lat, later, lateral	side, wide
line, lign	line
log, logy	word, study, reasoning
medi, mid	middle
meter, metr, metry	measure

Root	Meaning
mut, mute	change, exchange
neg	deny
ord, ordin, orn	put in order, arrange
pend	weigh, hang
phon	sound
phot, phos	light
plex, plic, ply, ple	fold, combine
poly	many
pon, pos	put, place
prim, prin, prior	first
radi, radic	root
reas, rati	think, calculate
reg, rect	rule, right, direct
scop	look
scrib, script	write
sec, sect, seg	cut
ser, sor, sort	attach to one another
sta, stat, sist, stit	stand, stay
spec, spic	look
tact, tang, ting, teg, tax	touch
tend, tens, tent	extend, stretch
term, termin	limit, end
tract	draw, drag, pull
vari	change
vers, vert	turn
vid, vis	see

Word Roots (continued)

Roots in Action

Let's see how recognizing roots can help you decipher unfamiliar words. Read the passage. Imagine that the underlined words are completely unknown to you. Then read the paragraphs below the passage, which show the kind of thought process a reader might go through to figure out what the underlined words mean.

> Biological life is 50 percent water. Almost three-quarters of the earth's surface is covered with water at an average depth of about 10,000 feet, so deep that people think the amount of water available in the universe is unlimited. It is not.

> Biological. *Bio* plus *logic* plus *-al*. Related to the study of life. Must mean the life that life scientists study.

Application

The following text has some words in it that you probably don't know. Use your knowledge of roots, along with any other vocabulary strategies you choose, to figure out their meanings.

> *Numbers and Operations*
>
> Two numbers are additive inverses or opposites of each other if their sum is zero. For example, 2 and –2 are additive inverses because –2 + 2 = 0.
>
> Two numbers are multiplicative inverses or reciprocals of each other if their product is 1. For example, 2 and $\frac{1}{2}$ are multiplicative inverses because $2 \times \frac{1}{2} = 1$.
>
> An operation is commutative if you can change the positions of the numbers involved without changing the result. For example, addition is commutative: 2 + 3 = 3 + 2. Subtraction is not commutative: 2 – 3 = –1, but 3 – 2 = 1. Multiplication is commutative: $2 \times 3 = 3 \times 2$. Division is not commutative: $\frac{2}{3}$ does not equal $\frac{3}{2}$.
>
> An operation is associative if you can change the order in which you group numbers together to operate on them without changing the result. For example, addition is associative: (2 + 3) + 4 = 5 + 4 = 9, and 2 + (3 + 4) = 2 + 7 = 9. (The parentheses indicate which pairs of numbers you add together first.) Subtraction is not associative: (2 – 3) – 4 = –1 – 4 = –5, but 2 – (3 – 4) = 2 – (– 1) = 3. Multiplication is associative: $(2 \times 3) \times 4 = 6 \times 4 = 24$, and $2 \times (3 \times 4) = 2 \times 12 = 24$. Division is not associative: $(2/3)/4 = (2/3) \times (1/4) = 2/12 = 1/6$, but $2/(3/4) = 2 \times (4/3) = 8/3$.
>
> Multiplication is distributive over addition and subtraction. For example, $2 \times (3 + 4) = (2 \times 3) + (2 \times 4)$.

Word Roots *(continued)*

1. List four words that appear in the preceding text that have the same suffix as *additive* (perhaps spelled slightly differently).

2. The suffix used in question 1 makes a word into what part of speech?

3. Give the meaning of the prefix and root of each of the following words.

Word	Prefix and Its Meaning	Root and Its Meaning
inverse		
opposite		
multiply		
product		
commutative		
divide		
associative		
subtract		

PART 2
Prereading

Lesson 4
Previewing

Reading to Learn

To succeed in school and at work and have time to enjoy life as well, you need to learn to read nonfiction efficiently. Reading efficiently means reading in such a way that you learn as much as possible in as little time as possible. The purpose of this program is to teach you how to do that.

Most people learn to read by reading fiction. When we read fiction, we simply open to the first page and begin. We relax and let the story carry us along. If it's a good story, we may almost forget that we are reading. We become almost hypnotized into believing, at least for the moment, that we are in the life of the characters, seeing what they see and hearing what they hear. And if you read a good novel and don't remember much about it, it doesn't matter, as long as you enjoyed yourself. In some ways, reading a good novel is like watching a movie.

What we've just described is an appropriate way to read fiction but an inefficient way to read nonfiction. To read nonfiction efficiently you must learn to read in a completely different way. When you read a nonfiction book or article, to simply open to the first page and begin is time-consuming and unproductive. To relax and let the story carry you along doesn't work if there's no story. If you forget that you're reading, you'll probably forget *what* you're reading. Instead of letting the text hypnotize you into believing something that isn't true, you need to stay alert and think critically about what the author is saying. And if you don't remember what you've read, you've wasted your time.

Reading nonfiction efficiently is completely unlike watching a movie. A movie starts at the beginning and runs on its own time to the end. When you read nonfiction, you can begin anywhere you want. You can skim through a whole book or an article in a few minutes, read a little bit from the middle, jump to the end, go back to the table of contents and look up something that interests you. If you want to, you can read the book from back to front. The wonderful thing about printed text is that, unlike movies, television, or recorded music, it is scannable. When you read fiction, if you look ahead, you risk spoiling the story by giving away the ending. When you read nonfiction, you're free to take whatever path through the text you choose at whatever pace you choose.

Previewing *(continued)*

The 4 Ps   There are four things you can do before you begin reading a nonfiction book or article. They are called the 4 Ps. The 4 Ps will make your reading more efficient. Once you get used to them, the 4 Ps will save you time. They'll enable you to read faster, understand what you read better, and remember more of it later. The 4 Ps are

- **Preview.** Scan the entire text. Find out as much as you can about what you're going to read without actually reading it.
- **Predict.** Based on what you saw during your preview, what do you think the text is about?
- **Prior Knowledge.** What do you already know about the subject of the text?
- **Purpose.** What can you expect to accomplish by reading this text?

You can use a 4-P chart like the one below to help you keep track of your thoughts. Eventually, you will do these steps in your head. You will practice using a chart like this as you work through the prereading lessons

4-P Chart

1. Preview	2. Predict	3. Prior Knowledge	4. Purpose

Why Preview?   If you want to read efficiently, before you read any nonfiction text, you must find out as much as you can about what it says without actually reading it. How can you find out what a text says without reading it? By previewing it.

When you preview a movie, you watch snippets of the film that give you an idea of what the movie is about. A preview helps you decide if you are likely to enjoy the movie. When you preview a nonfiction reading assignment—a textbook chapter, a magazine article, a report—you get the main idea of what the reading is about. Previewing an informational text helps you understand and retain the information in the text.

Previewing (continued)

Previewing To preview an informational text, follow these steps.

1. **Read the title.** The title often tells you the main idea of the reading.
2. **Scan the reading.** Look for highlighted or boxed words. Watch for heads and subheads that tell you what the next paragraphs are about.
3. **Look at the graphics.** These include charts, tables, graphs, diagrams, drawings, photos, and so on. Read the accompanying captions, too.
4. **Skim the reading.** When you skim, you do not read word for word. You look at the first and last sentences in a paragraph, for example. In a long reading assignment, such as a whole textbook chapter, read the first paragraph and the last paragraph.

Application Use the previewing steps to preview the following article. Fill in the preview section of the 4-P chart that follows the reading passage. Some information has been filled in to get you started. (Do *not* read the whole article yet.)

The Pythagorean Theorem

One of the best-known theorems in mathematics is named for Pythagoras, a Greek philosopher and mathematician who lived in the sixth century B.C.E.

How to Draw a Right Angle on the Ground

Ancient Greek and Egyptian surveyors used a rope loop knotted into 12 equal lengths as a drawing tool.

Pythagoras learned from Greek and Egyptian land surveyors a centuries-old method they used when they needed to draw a line at right angles to another line on the ground. They used a rope with the ends tied together, marked off by knots into 12 equal sections. They drove a stake into the ground on the line where they wanted the right angle to be. They drove another stake into the ground three rope sections away from the first stake, then looped the rope around the two stakes. Next, they found the point that is four rope sections away from the first stake and five rope sections away from the second stake. They drove a stake there so that the rope looped around all three stakes, as in the drawing below. This makes a triangle whose sides are 3, 4, and 5 rope sections long. If the rope sections are all of equal length, the angle at the first stake will be a right angle.

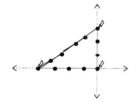

Stakes in the ground that form a 3-4-5 triangle automatically form a right triangle—no protractor is needed.

What Is It About 3, 4, and 5?
Pythagoras thought about this ancient technique and noticed that the numbers 3, 4, and 5 relate to each other in an interesting way. The sum of the

(continued)

Previewing (continued)

The Pythagorean Theorem (continued)

squares of the first two numbers equals the square of the third. As you probably know, the word square has a meaning in both geometry and arithmetic. In geometry, a square is a rectangle with all sides equal. A 3-unit square is a square whose sides are 3 units long. Pythagoras noticed that if you draw a square on each of the sides of a triangle like the one on page 18 whose sides are 3, 4, and 5 units long, the area of the square on the 3 side plus the area of the square on the 4 side equals the area of the square on the 5 side. You can verify this by counting the small squares in this drawing.

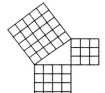

The area of the square drawn on the 5-unit side of a 3-4-5 triangle equals the sum of the areas of the other two sides.

In arithmetic, the square of a number is the number multiplied by itself. The square of 3, or "3 squared," is 3×3 or 9, the number of unit squares in a geometric square whose sides are 3 units long. Three squared is usually written 3^2. The superscript 2 means "take two 3s and multiply them together." So $3^2 + 4^2 = 9 + 16 = 25 = 5^2$.

What about Other Right Triangles?
Pythagoras asked the following question: Here we have an example of a right triangle where the area of the square on the side opposite the right angle (called the hypotenuse) equals the sum of the areas of the squares on the other two sides. Is this only true of 3-4-5 triangles? Are there other right triangles that have this property? Is it true of *all* right triangles?

Pythagoras proved that it is true of all right triangles. That's why the Pythagorean theorem is named after him. Since then, many proofs of the theorem have been discovered. Here's one.

Draw any two identical squares, A and B. Inside square A, draw a smaller square whose four corners touch square A, forming four identical right triangles. Inside square B, draw four right triangles identical to the four right triangles in square A, only this time tuck them into the corners of square B as shown in the drawing below. That leaves two smaller squares inside square B.

The smaller square inside square A is a square on the hypotenuse of each of the four right triangles. The two smaller squares inside square B are squares on the other two sides of the right triangles. Since squares A and B are identical and the right triangles are identical, the area of the smaller square inside square A must equal the sum of the areas of the two smaller squares inside square B.

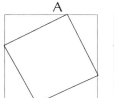

The square inside square A equals the sum of the two small squares inside square B.

Previewing *(continued)*

The questions below are the kinds of things you might ask yourself when you preview. Use them to help you record the information you gathered while previewing. Some information has been filled in to get you started.

- What does the title suggest the reading is about?
- What do the subheads tell you?
- What are the graphic elements about?
- From the sentences you skimmed, what seem to be some important concepts in the reading?

4-P Chart

1. Preview	2. Predict	3. Prior Knowledge	4. Purpose
Pythagorean theorem Subheads about triangles, certain numbers, right angles Pythagoras was a mathematician; asked questions about triangles.			

Lesson 5
Predicting

Predicting At the end of Lesson 4, you previewed the article on the Pythagorean theorem. The next step in the prereading process—the second P—is **predict.** Predicting in this context means asking and answering this question: Based on what you saw during your preview, what do you think the text is about?

When you preview a text, you grab most of the information in it in a small fraction of the time it would take you to read it. But if you preview a text without predicting, the information you gathered during your preview will disappear from your memory. To store it in your memory and be able to retrieve it later, you need to organize it.

Application Fill in your predictions about the Pythagorean theorem article in the 4-P chart below. Some information has been filled in to get you started. Change anything that is already there if you do not agree with it.

4-P Chart

1. Preview	2. Predict	3. Prior Knowledge	4. Purpose
	Will explain Pythagorean theorem, whatever that is. What is the round figure in the section about drawing a right angle? A clock of some kind? No. It's a rope used to measure for drawing—probably a right angle. The next paragraph tells how to use it. The next graphic is about counting squares—this is probably about figuring out the area.		

© 2002 J. Weston Walch, Publisher

LESSON 6
Prior Knowledge

In Lessons 4 and 5, you previewed the article on the Pythagorean theorem and predicted what information you would get from it when you actually read it. Now you're ready for the third P: **prior knowledge.** This is the information you already have about a subject.

Think of your memory as a library. In most libraries, the books are stored by subject matter. If you go to the math section, you'll find that the geometry books are on one shelf and the algebra books are on another. Storing books by subject matter makes it easy to find the book you need. If a math book was put away in the paleontology section, it might be lost forever.

Your memory works that way, too. When you get some new information from reading, you want to store it in the same place in your mind as your prior knowledge on the same subject. If you do, you'll be able to find it again easily when you need it.

Activating Prior Knowledge How do you store new information you get from reading where you can find it? You spend a few minutes recalling what you already know about the subject. That's the mental equivalent of going to the library shelf where that information belongs. What does the information you got from previewing remind you of? Make notes of whatever comes to mind.

Application Look over the information you put in the first two columns of your practice 4-P chart in Lessons 4 and 5. Then fill in the third column with the prior knowledge you have about what you noted.

4-P Chart

1. Preview	2. Predict	3. Prior Knowledge	4. Purpose

Lesson 7
Purpose

Reading with Purpose

Your **purpose** in reading a text is the reason you are reading it. It's what you need or want to get from reading. In reading, as in the rest of life, if you keep your purpose clearly in mind and let it guide what you do, you'll probably succeed in it.

What is your purpose in reading the Pythagorean theorem article? Clearly, your purpose has been to learn the prereading techniques. What purpose might someone have for reading the article outside the context of this book? Here are some possibilities. You can probably think of more.

- Gather information for a study of the development of mathematics in ancient times.
- Gather information for a paper on Pythagoras.
- Learn about the Pythagorean theorem so as to be able to use it to solve geometry problems.

Application

If you were reading the Pythagorean Theorem article outside the context of this program, what do you think your purpose in reading it might be? Write it in the chart below.

4-P Chart

1. Preview	2. Predict	3. Prior Knowledge	4. Purpose

Now read the entire article about the Pythagorean theorem. Then answer the questions below.

1. Was the reading pretty much what you expected, or were there some surprises?

2. Do you think using the four prereading strategies helped you understand what you read? Explain.

PART 3
Reading Strategies

LESSON 8
Introduction to Reading Strategies

In this part of the book, you will learn five strategies you can use to make your nonfiction reading more efficient. And, as you know, more efficient reading means you learn more with less time and effort.

Why five strategies? Why not just learn the best one and use it all the time? There is no "best one." Texts are all different, and readers are all different. Some reading strategies are better suited to some texts than to others. And different readers like different strategies. In the following lessons, you will learn and practice five different reading strategies. After that, it will be up to you to choose which one to use when you read. The important thing is not to use the "right" strategy the "right" way; the important thing is to read efficiently. Try all the strategies. Then do what works for you.

Five Reading Strategies

Here are the five strategies you will learn.

- **KWL.** KWL stands for "What I **K**now, What I **W**ant to Know, What I **L**earned." You take notes on a three-column form while prereading and reading.
- **SQ3R.** SQ3R stands for **S**urvey, **Q**uestion, **R**ead, **R**ecall, **R**eflect, a five-step process. The Survey and Question steps are similar to the prereading techniques you learned in Lessons 4 through 7, the 4 Ps.
- **Semantic Web.** During prereading and reading, you draw a diagram that maps the information in a text, showing how various concepts and facts relate to each other. *Semantic* means "about meaning," so a semantic web is a kind of meaning map.
- **Outline.** An outline is a way of noting the information gathered during prereading and reading, listing main ideas with supporting ideas and details indented under them.
- **Structured Notes.** Preview the text, see what its semantic structure is, draw your own graphic organizer, and then fill it in.

Lesson 9
KWL

KWL

In this lesson, you'll learn a reading strategy called KWL. When you use KWL, you use a three-column form to structure your notetaking.

The first column is headed **K,** for What I Know. As you preread you make notes in this column on what you get from previewing and recalling prior knowledge you may have about the subject of the text.

The second column is headed **W,** for What I Want to Know. As you preread, you make notes in this column, usually in the form of questions, on what you want and expect to find out that will fulfill your purpose in reading the text.

The third column is headed **L,** for What I Learned. You make notes in this column as you read and after you finish reading. If you get the information you expected and wanted to get—and most of the time you will get at least some of it—you can note the answers next to the questions in the second column. Sometimes you'll get something different from what you expected. You'll be disappointed or pleasantly surprised by what you read. Make notes on that, too, in the third column.

K What I KNOW	W What I WANT to Know	L What I LEARNED

KWL in Action

Use this reading passage to try out the KWL strategy. First, preview the text. There are no illustrations or other graphic elements, but we find out a lot by simply reading the title and the first sentence.

> *How the Ancient Greeks Knew the Earth Was Round*
>
> Centuries before the voyages of Columbus and Magellan, the scientists of ancient Greece suspected that the earth was round rather than flat, even though they could not prove it. They had three reasons.

KWL (continued)

Applying the KWL reading strategy to this text, you might start by asking yourself how *you* know that the earth is round and writing that information in the K column. For example, you have probably heard or read about ships, airplanes, and satellites that have circled the earth. Second, you might have heard that when a ship is a long way off on the ocean, you can only see the top of its superstructure; the bottom part is hidden by the curvature of the earth. If that is what you knew before reading this text, then your KWL chart might look like this.

KWL Chart

K What I KNOW	W What I WANT to Know	L What I LEARNED
• Ships, planes, satellites circle Earth • Bottom of ships below horizon	• Greeks' reasons • Did Greeks know about this?	

You may want to add your predictions to the W column, perhaps turning statements into questions.

Now you are ready to read the text.

How the Ancient Greeks Knew the Earth Was Round

Centuries before the voyages of Columbus and Magellan, the scientists of ancient Greece suspected that the earth was round rather than flat, even though they could not prove it. They had three reasons.

First, they observed that during eclipses of the moon, the part of the moon that was darkened was always round. They reasoned that lunar eclipses were caused by the earth coming between the sun and the moon and that the earth must be spherical. If the earth were a flat disc, as people had always believed, it would sometimes cast a shadow that was elongated and elliptical.

Second, the Greeks were traders and traveled all over the eastern Mediterranean Sea. They knew that the North Star (which we now know is over the North Pole) and other stars near it appear lower in the sky when seen from Egypt than when seen from northern Greece. They even attempted to estimate the circumference of the earth based on their observations. Unfortunately, though, because they did not have an accurate way to measure long distances on the surface of the earth, their estimates were not correct.

Third, the ancient Greeks knew that as a ship comes to port, those on the shore see the sails coming over the horizon first and only later see the hull.

Adapted from Stephen W. Hawking, *A Brief History of Time: From the Big Bang to Black Holes.* New York: Bantam Books, 1988.

KWL (continued)

Now you can fill in the L column.

K What I KNOW	W What I WANT to Know	L What I LEARNED
• Ships, planes, satellites circle Earth	• Greeks' reasons	1. Earth's shadow on the moon during eclipse is round. 2. North Star seen at different heights in Egypt (south) and Greece (north)
• Bottom of ships below horizon	• Did Greeks know about this?	3. Yes; Reason #3

The result is a concise, usable set of notes on the text.

Application

Make notes on the following text using the KWL reading strategy. Then answer the questions that follow the reading.

Time Zones in the United States

The distance around the world (its circumference) is approximately 24,000 miles at the equator. There are 24 hours in a day, and 360° in a circle—like the Equator. If we divide the Equator into 24 equal parts, each part will take up 15° (360° ÷ 24 = 15°). These 24 areas are called **time zones**. Because 24,000 ÷ 24 = 1000, each time zone is 1000 miles wide at the Equator. As we move north or south from the Equator, each time zone becomes narrower, as shown in Figure 1. However, each zone is still 15° wide. As you can see from Figure 1, the distance around lines of latitude becomes smaller as we move toward the North or South Pole along lines of longitude.

Lines of latitude across the United States are from approximately 25° to 50°. The lines of longitude range from about 70° to 125°—a total of about 55° in the east-west direction. If we divide 55° by 15° we get about $3\frac{2}{3}$ time zones across the United States. The average width of the United States is approximately 2800 miles.

As a result, the country is divided into four time zones—Eastern, Central,

(continued)

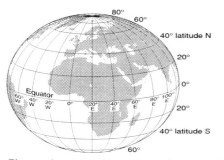

Figure 1. Lines of latitude circle the earth. They are parallel to the Equator, which is 0° latitude. Lines of longitude run from the North Pole (90° north latitude) to the South Pole (90° south latitude).

KWL (continued)

Time Zones in the United States (continued)

Mountain, and Pacific. These are abbreviated as EST, CST, MST, and PST for Eastern Standard Time, Central Standard Time, and so forth. As you can see from Figure 2, the boundaries of these time zones do not always lie along lines of longitude. This is because political and social factors enter into deciding on time zones.

During the summer months, many states change to daylight saving time—DST. This means that in the spring (usually in April), they move the time ahead one hour. In the fall (usually October), they resume standard time and turn their clocks back one hour.

Dividing the nation's average width, 2800 miles, into four zones gives an average width of only 700 miles for each zone. This is $7/10$ of the average width of a time zone at the equator. At the poles, the lines of longitude come together, and the width of each zone shrinks to zero.

Anyone traveling across the country or making phone calls to people in other time zones must take the difference in time into account. No one wants to be called after they have gone to sleep.

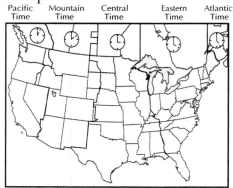

Figure 2. The continental United States has four major time zones.

Robert Gardner and Edward A. Shore, *Middle School Math You Really Need*, J. Weston Walch, Publisher, 1997.

K What I KNOW	W What I WANT to Know	L What I LEARNED

KWL (continued)

QUIZ: Time Zones in the United States

Circle the letter of the correct answer.

1. Between San Francisco and New York there is a small difference in _____ but a big difference in _____.
   (a) longitude, latitude
   (b) latitude, longitude
   (c) longitude, elevation
   (d) elevation, latitude
   (e) latitude, elevation

2. If the president makes a speech to the nation that is broadcast live from the White House in Washington, D.C. at 9:00 P.M. EST, at what time will it be viewed in Colorado (Mountain Time)?
   (a) 11:00 P.M.
   (b) 10:00 P.M.
   (c) 9:00 P.M.
   (d) 8:00 P.M.
   (e) 7:00 P.M.

3. The 48 states of the continental United States (not including Alaska and Hawaii) span approximately _____ degrees of latitude.
   (a) 25
   (b) 50
   (c) 70
   (d) 125

4. The boundaries of time zones do not always coincide with lines of longitude because sometimes they follow
   (a) major highways
   (b) mountain ranges
   (c) property lines
   (d) state and county lines

5. It has been suggested that time zone boundaries should be changed so that the entire continental United States would all be on the same time. One advantage would be that people in all parts of the country would be at work at about the same time. What problems would this change cause?

Lesson 10
SQ3R

SQ3R

In this lesson, you'll learn about a reading strategy called SQ3R. SQ3R is somewhat similar to KWL but more sophisticated. Like KWL, SQ3R is comprehensive in that it includes prereading, reading, and postreading phases.

SQ3R stands for **S**urvey, **Q**uestion, **R**ead, **R**ecall, **R**eflect. The Survey and Question steps of SQ3R are the prereading steps. When you do them, you do much the same kinds of things as when you do the 4 Ps of prereading. The Review and Reflect steps are similar to the What I Learned step of KWL. Here are the five steps of SQ3R.

- **Survey.** This step combines the Preview and Prior Knowledge prereading steps you learned earlier. The goal is to find out as much as possible about the text without actually reading it. You do that by surveying headings, illustrations, captions, introductory or summary material, first and last sentences of paragraphs, and so on. At the same time, you think about what you already know about the subject.

- **Question.** This step more or less combines the Predict and Purpose steps of the 4 Ps of prereading. You write the questions that express what you want and expect to find in the text.

- **Read.** This is the step where you actually read the text, looking for the answers to your questions.

- **Recall.** In this step, you write the answers to your questions and make notes on any information you found that was not anticipated by your questions.

- **Reflect.** In this step, you think about what you have read in the light of your prior knowledge and your purpose in reading. Did you get what you wanted and expected from the text? Is the text clear? Did this reading lead to any more questions you want answered?

SQ3R in Action

You will practice the SQ3R strategy using the text on page 34. But first, a warning about titles. When you read nonfiction, you must preread. When you preread a piece, you must pay close attention to its title. The title of a piece usually tells you clearly and concisely what it is about—but not always. In the

SQ3R (continued)

last lesson, you read "How the Ancient Greeks Knew the Earth Was Round" and "Time Zones in the United States." The titles of those pieces clearly identified their subject matter. The title of the text on page 34, on the other hand, is misleading. It suggests that if you read the text you will learn how to get rich in 30 days. You won't. The title is intended to get your attention—which it does. But it does not help you get meaning from the text. Remember, titles don't always mean what they say.

Now apply the SQ3R strategy. Start by surveying the text.

Survey The title and the first sentence are not helpful, but we notice in the table that as the day increases by one the charge increases very rapidly. In the last paragraph, we see that the words *exponential growth* are italicized.

SQ3R

S Survey	Q Question	R Read	R Recall	R Reflect
table shows as day increases, charge increases rapidly exponential growth				

Question As a result of your survey, you might ask questions like those shown in the chart.

SQ3R

S Survey	Q Question	R Read	R Recall	R Reflect
	Why does the charge increase so fast? Does exponential growth mean very rapid growth? I remember hearing about exponents in math class. Is exponential growth related to that?			

SQ3R (continued)

Read Now read the text. Look for the answers to your Q questions.

Get Rich in Just 30 Days

Think of a chore that needs to be done every day, like sweeping the floor or washing the dishes. Make the following offer to the person usually responsible for the job: "I'll do this job every day for a month. I'll only charge one cent for the first day, two cents for the second day, four cents for the third day, and so on. The charge for each day will be twice as much as the day before. Okay?"

Day	Charge	Day	Charge
1	1¢	5	16¢
2	2¢	6	32¢
3	4¢	7	64¢
4	8¢	8	128¢

Doesn't sound like you'd make much, does it? Wrong. Figure it out. If you actually got someone to pay you every day for a month according to this deal, you'd end up with an amount that could be the grand prize in a lottery. The only problem would be finding someone who could afford to pay you.

Even if you can't really get rich using this scheme, it's a fun way to illustrate the mathematical concept of *exponential growth*. If something grows exponentially, it grows really fast. Why? Because the bigger it gets, the faster it grows.

Adapted from Marilyn Burns, *The I Hate Math! Book.* New York: Little Brown and Company and Yolla Bolly Press, 1975.

Now your chart might look like this.

SQ3R

S Survey	Q Question	R Read	R Recall	R Reflect
	Why does the charge increase so fast?	It doubles every day. Each day a larger number gets doubled.		
	Does exponential growth mean very rapid growth?	Yes.		
	I remember hearing about exponents in math class. Is exponential growth related to that?	Yes.		

Recall In this section, note additional information from the reading not anticipated by your questions.

Reflect How you respond to a text in the Reflect step depends partly on your prior knowledge. Someone who had never studied exponents might say, "This text gives an example of exponential growth, but it does not really explain what

SQ3R (continued)

exponential growth is." Someone who had studied exponents and already understood what exponential growth is would understand the text better. Such a reader would understand that the charge for doing dishes on day n is 2^n¢. He or she would be able to calculate, if they wanted to take the trouble, that after two weeks, on day 15, the charge would be 2^{15}¢ or $327.68. After a month, on day 31, the charge would be 2^{31}¢ or $21,474,836.48.

Application Try applying the SQ3R reading strategy to this text. On page 35, fill in the first two columns using prereading strategies. Then read the text and complete the remaining columns. Finally, answer the questions that follow the reading.

The Geometry of Honeycombs

Bees' honeycombs, the containers where they store their honey, are tessellated regular hexagons made of wax. You'll recall that a hexagon is a six-sided polygon. *Regular* means all sides and angles are equal. *Tessellated* means the regular hexagons fit together with no gaps and no overlapping. Each corner point is surrounded by exactly three hexagons. There is no wasted space between cells.

Is this honeycomb design the most efficient one the bees could have used? Does it provide for storage of the greatest payload of honey for the least cost in wax? Or would some other regular polygon work better?

There are only three regular polygons that tessellate: triangles, squares, and hexagons. What if bees used triangles? Regular (equilateral) triangles tessellate nicely.

But you can see from the drawing that triangular honeycombs use much more wax to store the same amount of honey than hexagons do. Look at any point where six triangles come together and imagine taking away all the line segments that meet at that point. What's left? A hexagon.

What about using squares? Squares (regular quadrilaterals) tessellate.

But are squares more efficient than hexagons? It's hard to tell from the drawing. We'll have to calculate. If the area of the polygon used is the measure of how much honey you can store in it, and if the perimeter is the measure of how much wax it takes to build it, then the ratio of area to perimeter is the measure of how efficient each polygon is. Let's calculate that, using the formulas for area and perimeter. We'll assume that every side of every polygon is one unit long. We'll round off to the nearest hundredth.

Polygon	Side	Area (square units)	Perimeter	Area/Perimeter
Triangle	1	0.43	3	0.14
Square	1	1.00	4	0.25
Hexagon	1	2.59	6	0.43

The hexagon is the clear winner! How do the bees know that?

SQ3R *(continued)*

S Survey	Q Question	R Read	R Recall	R Reflect

SQ3R (continued)

QUIZ: The Geometry of Honeycombs

Circle the letter of the best answer.

1. A polygon is
   (a) a closed geometric figure made of line segments.
   (b) a closed geometric figure made of line segments whose sides and angles are equal.
   (c) a closed geometric figure made of line segments whose sides and angles are equal and that fills up the plane without gaps or overlapping.
   (d) a departed parrot.

2. Tessellated polygons are a more efficient design for honeycombs because
   (a) bees prefer them.
   (b) they have equal sides and angles.
   (c) they have no wasted space between cells.
   (d) they have six sides.

3. In the context of this text, *efficient* means
   (a) catching the most fish with the least bait.
   (b) learning the most with the least time and effort.
   (c) producing the most work with the least energy.
   (d) storing the most honey with the least wax.

4. Tessellated regular hexagons are a better design for honeycombs than the others considered in this text because
   (a) their ratio of area to perimeter is the greatest.
   (b) their ratio of perimeter to area is the greatest.
   (c) they have the most sides.
   (d) triangles require more wax.

5. The text ends with "The hexagon is the clear winner! How do the bees know that?" What's your answer to that question?

Lesson 11
Semantic Web

Semantic Web

The reading strategies you have already practiced—KWL and SQ3R—suggest that you organize your reading notes according to when you take them, that is, with your prereading notes in one place, your reading notes in another, and your thoughts after reading in another.

The reading strategies we will show you in this lesson and in Lessons 12 and 13—semantic web, outline, and structured notes—take a different approach. These strategies suggest that you organize your notes—prereading, reading, and postreading—entirely by subject matter, with related ideas together. They offer you three different ways to do that.

In this lesson, we will show you the reading strategy called semantic web. The word *semantic* means "about meaning." A web—a spider's web or the World Wide Web—is a set of connections among points. A semantic web is a sort of map of the meaning of a text with closely related facts and ideas close together and connections between them shown by lines.

Semantic Web in Action

Watch how a semantic web can be made based on this paragraph by Bertrand Russell, the great British mathematician and philosopher. Follow along with the thought process of one reader.

> *Approximation*
>
> Although this may seem a paradox, all exact science is dominated by the idea of approximation. When a man tells you that he knows the exact truth about anything, you are safe in inferring that he is an in-exact man. Every careful measurement in science is always given with the probable error, which is a technical term, conveying a precise meaning. It means: that amount of error which is just as likely to be greater than the actual error as to be less.
>
> From Bertrand Russell, *The Scientific Outlook*. New York: W.W. Norton and Company, 1931.

In previewing this paragraph, a reader might begin a semantic web with notes like these: I haven't come across the word *approximation* before, but it reminds me of the word *approximately*, which means "close but not exactly." They're the same root with different suffixes. So I'll note that with a line to *approximation*.

Semantic Web (continued)

The word *paradox* appears in the first sentence. I know that a paradox is something that seems like it shouldn't be true but is. I'll need to find out what the paradox is. I'll put that on the web.

The word *error* appears three times in the paragraph. It must be important. I'll put that on the web.

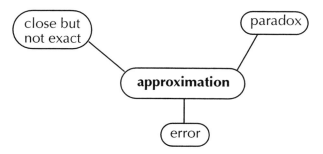

Now I'll really read the paragraph and fill in the semantic web. The first sentence says that the paradox is that exact science that is dominated by approximation, which means "close but not exact." Seems like a contradiction.

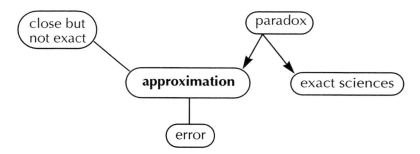

The second sentence says that if someone says they know the "exact truth" about anything then they are not exact. More paradox.

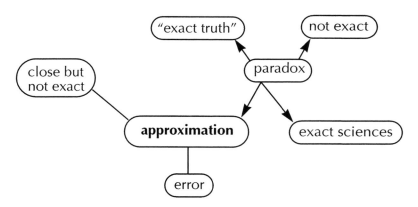

Semantic Web (continued)

The next sentence says that careful scientific measurement is always given with the probable error. What does that mean? The next sentence explains it. It means that a careful scientist knows that measurements won't be exactly correct. They are bound to have some amount of error in them. So the careful scientist tries to estimate how big the error probably is. The probable error is "that amount of error which is just as likely to be greater than the actual error as to be less." To save space on the semantic web, I'll use the mathematical symbols for greater than (>) and less than (<).

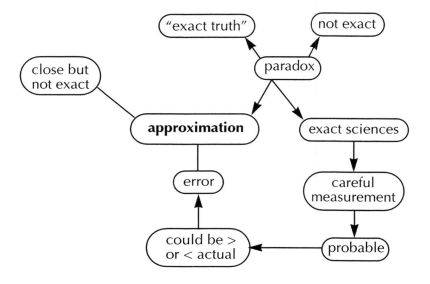

Application

As you read the article below, fill in the semantic web that follows. The web has been started for you, but add or remove lines and circles as needed. Then complete the quiz.

What Is Infinity?

Part of mathematics is thinking about sets of things and about counting the things in sets. A set is simply a collection of things that is well enough defined so that it is always clear whether a given thing is a member of the set or not. As every small child knows, when we count the members of a set, we map the set of counting numbers onto them. Starting with 1 and taking the counting numbers in order, we assign a counting number to each member of the set: "1, 2, 3, 4, ..." The greatest counting number assigned is the count of the set.

Mathematicians use the term *proper subset* for a set that contains some but not all of the members of another set. For example, the set of all girls in a coeducational school is a proper subset of the set of all students in the school. A proper subset of a finite set always has fewer members in it than the set of which it is a part. If there are boys in

(continued)

Semantic Web (continued)

What Is Infinity? (continued)

a school, then the set of all girls must have fewer members than the set of all students.

Since ancient times, people have speculated about infinite sets, but until fairly recently, no mathematician had clearly defined what infinite meant. In the late nineteenth century, Georg Cantor, a Russian-born mathematician living in Germany, came up with a beautifully simple and clear definition. He said that a set is infinite if it contains a proper subset that has just as many members in it as the set itself has.

For example, the set of even numbers is a proper subset of the set of counting numbers. All even numbers are counting numbers, but not all counting numbers are even numbers.

On the other hand, there are just as many even numbers as there are counting numbers. You could map the set of counting numbers onto the set of even numbers and never run out of even numbers. Every counting number maps onto its even double, as in the table below. Therefore, the set of counting numbers is infinite.

Counting numbers	1	2	3	4	5	6	7	8	9	10	11	...
	↕	↕	↕	↕	↕	↕	↕	↕	↕	↕	↕	
Even numbers	2	4	6	8	10	12	14	16	18	20	22	...

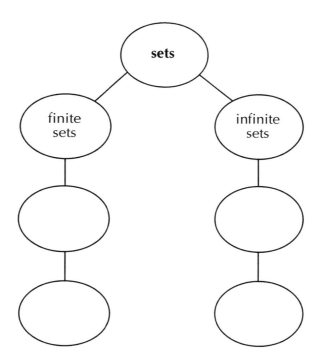

Semantic Web (continued)

QUIZ: What Is Infinity?

Circle the letter of the best answer.

1. The word *map* as it is used in this text could be applied to which of the following situations?
    (a) Central High School is a coeducational school.
    (b) Everyone please choose a partner for the next dance.
    (c) Mary drove 5 miles south on Route 89.
    (d) Some counting numbers are even and some are odd.

2. A set is infinite if
    (a) it contains a proper subset that has fewer members in it than the set itself.
    (b) it contains a proper subset that has more members in it than the set itself.
    (c) it contains a proper subset that has the same number of members as the set itself.
    (d) it contains a subset that has the same number of members as the set itself.

3. The set of _____ is a proper subset of the set of _____.
    (a) all pine trees, all palm trees
    (b) all pine trees, all trees
    (c) all trees, all pine trees
    (d) all trees, all trees

4. Which of these sets is infinite?
    (a) The set of all multiples of 3
    (b) The set of all fractions whose denominator is 3
    (c) Both of the above
    (d) Neither of the above

5. In your opinion, is the set of all subatomic particles (electrons, protons, neutrons, and so forth) in the universe infinite? Why or why not?

Lesson 12
Outline

Why Outline?

Outlining is another reading strategy that suggests organizing your reading notes according to subject matter. Like a semantic web and structured notes, an outline does not tell you how to distinguish your prereading and postreading notes—your prior knowledge, your expectations, your opinions and evaluations—from your reading notes. You must find your own way to do that. For example, you might put your own thoughts in brackets [ ]. It doesn't matter how you do it, as long as you are consistent about it so that when you come back to your notes later you can tell who said what.

Outlining forces you to distinguish main ideas from supporting ideas and details. This can be both good and bad. It's good if the text you are reading states its main ideas clearly and follows them with subordinate ideas and details. But not all texts are written that way. Some texts begin with details and end with a main idea; such a text may be difficult to outline.

The strength of outlining is that it supports you in storing information in your notes in the same order that it is presented in the text. For this reason, outlining lends itself well to taking notes on a narrative because it preserves the order of events.

Outline in Action

By custom, the top-level headings of an outline are numbered with Roman numerals; second level, capital letters; third level, Arabic numerals; fourth level, lowercase letters; fifth level (if there is one), lowercase Roman numerals. We'll demonstrate outlining using this text.

> *Pythagoras*
>
> Pythagoras of Samos, for whom the Pythagorean theorem is named, lived in Greece from 569 to 475 B.C.E. He is often described as the world's first pure mathematician because he was the first to discuss mathematics as a coherent intellectual discipline in its own right rather than a collection of rules of thumb for solving practical problems.
>
> Pythagoras was a philosopher and religious leader. He and his followers studied a wide variety of subjects including architecture, astronomy, athletics, dance, medicine, music, nutrition, politics, and ethics.
>
> Pythagoras taught that numbers, which are both concrete and abstract, are the ultimate substance underlying all things. For example, the number 1,
>
> *(continued)*

Outline (continued)

Pythagoras (continued)

which is physical as in counting, can stand for the unity of all creation. An accomplished musician, Pythagoras discovered the numerical relationships between musical pitches, which can evoke human emotions, and the lengths of vibrating strings.

The Pythagoreans, as he and his group were called, thought of mathematics as the bridge between the visible and invisible worlds: not only a useful tool for trading goods and building houses in the physical world, but also a way to escape from the transitory and misleading world of the senses to a realm of universal and unchanging truths. Pythagoras taught that focusing on mathematics calms and purifies the mind and, through disciplined effort, can lead to true happiness.

This is how a reader might outline the reading.

I. Pythagoras
   A. Theorem [hypotenuse of right triangle. I still don't get it.]
   B. Greece, 569–475 B.C.E.
   C. World's first "pure" mathematician
      1. Math as a discipline vs. practical
   D. Philosopher & religious leader
      1. Studied: many subjects: astronomy, ethics
   E. Numbers the ultimate substance underlying everything
      1. 1 = unity of creation
      2. Musical pitches ↔ length of vibrating strings
   F. Math
      1. Bridge visible → invisible
      2. Escape world of senses to universal truths
      3. Calms, purifies mind [sometimes]
      4. Leads to true happiness [How long do I have to wait?]

Outline (continued)

Application Read this text and make notes using the outlining strategy. You may use the blank outline that follows the reading, changing it as needed. Or you may choose to make a completely original outline. After you have completed your reading and your outline, take the quiz.

Universal Time

Universal Time (UT) is the 24-hour clock version of Greenwich Mean Time (GMT), the standard time of Great Britain, where the Greenwich Meridian, 0° longitude, is located. Time measurements in the earth sciences ordinarily use UT, as do astronomical tables.

Universal Time and all other standard times are forms of solar time in that they measure time by the changing relationship of the Sun to the meridian at a certain place on the Earth. The meridian of a place is the arc of the great circle that runs from the north point of the horizon to the zenith, directly overhead, to the south point of the horizon. At northern latitudes, when the Sun crosses the meridian, it is due south of the observer. In old books, the crossing of the meridian is called the Sun's "southing" and defines local noon.

The trouble with using the real Sun as the basis of timekeeping is that because of the elliptical shape of the Earth's orbit around the Sun, the Sun's apparent motion is not constant. Time measurements using the southing of the Sun can vary by as much as 15 minutes from measurements that use a constant clock. This became a problem with the invention of the mechanical clock, which was constant and could not easily be adjusted to fit the Sun's variations. So astronomers created a fictitious "mean Sun," which moves along the celestial equator at a constant rate. Noon local mean time then, is the southing of the mean Sun, which moves at the average rate of the true Sun and coincides with it twice a year.

In theory, every place on Earth could keep its own local mean time based on the angle between the local meridian and the mean Sun. In practice, the Earth has been divided into 24 time zones. Every place keeps the time of a nearby meridian of longitude, which is usually a multiple of 15° east or west of Greenwich. Most of these time zones are a whole number of hours ahead [of] or behind Greenwich.

From R. Hand and J. Brackett, "How to Cast a Natal Horiscope" as found in Neil F. Michelsen, *The American Book of Tables*. San Diego: ACS Publications, Inc.

Outline *(continued)*

I. _____

    A. _____

        1. _____

        2. _____

        3. _____

    B. _____

        1. _____

        2. _____

        3. _____

II. _____

    A. _____

    B. _____

    C. _____

Outline (continued)

QUIZ: Universal Time

Circle the letter of the best answer.

1. At 1500 UT, what time is it in London?
   (a) 2:00 P.M.
   (b) 3:00 P.M.
   (c) 4:00 P.M.
   (d) 5:00 P.M.

2. If you were in South Africa when the sun crossed the meridian, in what part of the sky would you see it?
   (a) north
   (b) south
   (c) east
   (d) west

3. If you were in South Africa when the sun crossed the meridian, what time of day would it be?
   (a) morning
   (b) midday
   (c) afternoon
   (d) evening

4. Time measurements using the southing of the sun can vary by as much as 15 minutes from measurements that use a constant clock. Why?
   (a) Some clocks are inaccurate.
   (b) The rotation of the earth speeds up and slows down.
   (c) The sun goes south in the winter.
   (d) The sun seems to speed up and slow down.

5. In theory, every place on Earth could keep its own local mean time based on the angle between the local meridian and the mean sun. What problems would arise if we kept time this way?

LESSON 13
Structured Notes

Choosing Structured Notes

The reading strategies in this book use graphic organizers to help you to understand and to remember what you read. The 4 Ps, KWL, and SQ3R each provide you with a graphic form to fill in as you read. A semantic web suggests that you represent the semantic structure of the text visually by drawing a diagram of it. Outlining suggests that you do so by indenting.

In this lesson, you'll learn about a flexible reading strategy called structured notes. This strategy does not provide you with a form to fill in. It asks you to design a form yourself for the particular text you are reading. The structured notetaking strategy is to preview the text, see what its semantic structure is, draw your own graphic organizer, and then fill it in.

Organization

Here are some patterns of organization that writers commonly use and some suggestions for the kinds of graphic organizers you might create to take notes on texts that follow those patterns. The patterns are divided into four groups according to the general principle that governs the order of sentences, paragraphs, and groups of paragraphs. The four groups are **time, list, visual,** and **logical.**

Time. Many texts are organized by time. They are narratives—this happened, and then this happened, and then this happened—or descriptions of the steps in a process—first do this, then do this, then do this. Many texts, especially magazine articles, begin with a story paragraph to capture your interest ("When Julia Gomez went to her school locker on the morning of April 17, everything seemed normal . . . "), then switch to another pattern of organization. ("What happened to Julia Gomez is typical of what happens to thousands of middle school students every day . . . "). So don't assume that, just because a text begins with a narrative, it will continue as a narrative until the end. Preview it.

When you take notes on a text that is organized by time, the important thing is to make sure that the time order is preserved in your notes. Outlining does that. So does a flowchart. There are lots of ways to draw a flowchart.

Structured Notes (continued)

Here's one way. The arrows show clearly what comes first, second, third, and so on.

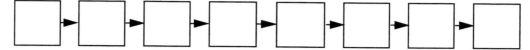

List. Writers often write about subjects that, unlike a story or process, don't have a natural order for sentences and paragraphs to fall into. They are lists. The order is arbitrary. Here are some examples of the kinds of texts that are essentially lists.

- The Causes of World War I
- Birds of the Middle West
- Three Reasons Why You Should Vote for Proposition 5
- Objects Found at the Crime Scene
- The Consequences of Global Warming

In general, outlining is a good way to take notes on a list text. Outlining is especially appropriate for handling lists of lists. A semantic web may also work well for a shorter list text.

There are two special kinds of list patterns that call for slightly different graphic organizers. One is comparison and contrast; the other is main idea and details. A **compare-and-contrast** text talks about two (or more) things and lists the ways in which they are similar and the ways in which they are different. For example,

- Mathematics Education in the United States and Japan
- How Boys and Girls Talk to Each Other
- Word Processing under Windows and Macintosh

A good graphic organizer to use for a compare-and-contrast text is two overlapping circles (a Venn diagram). Each of the two things being compared and contrasted has its own circle. The similarities go inside the area where the circles overlap. The differences go outside it.

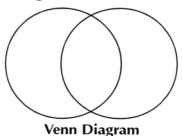

Venn Diagram

Structured Notes *(continued)*

Main idea and details is a common pattern of organization for paragraphs and longer texts. Typically, the main idea is expressed in a general statement of a sentence or two. This is followed by a list (or a time sequence) of more specific statements. Without the specifics, the general statement would be vague or dubious. Without the general statement, the specifics would seem pointless. Together, both are clear and credible. The main idea does not have to come at the beginning of a paragraph; it may come at the end or even in the middle. A semantic web or an outline works well as a graphic organizer for a main-idea-and-details paragraph. For a longer text on this pattern, you may want to draw something like this.

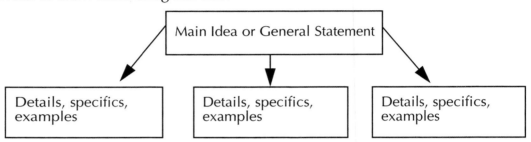

Visual. Sometimes a text describes or explains something visual. Typically, text of this kind is accompanied by visual information—maps, photographs, illustrations, graphs, tables, diagrams, or the like—and the organization of the text follows the organization of the visual information or of the subject itself. For example, a text describing a building might describe the outside first and then the inside. The description of the outside might start with the bottom and move to the top, or the other way around. The description of the inside might follow the path that a typical visitor might take through the building.

The important thing to keep in mind when you read a visual text is that the visual information is often more important than the words. Therefore, you must find a way to get the visual information into your reading notes. If you can include a photocopy of the visual information in your notes, that's ideal, but you can't always do that. Often a sketch of the most important photo, graph, or diagram is good enough, and the act of sketching will help you remember. You can tie your notes to the sketch with callouts. (A *callout* is text connected to a visual element by a line.) Here's an example of a sketch with callouts.

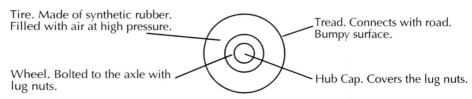

Structured Notes (continued)

Logical. Logical is a catchall category for texts that aren't clearly organized by a time, list, or visual pattern. But logically organized texts do have some characteristics in common. In general, they begin with what the reader knows, understands, believes, or values, and proceed by some sort of reasoning to what the reader did not previously know, understand, believe, or value. Logically organized texts tend to say, "We know that this is true. Therefore, this is true and this is true. Therefore, this is true."

The classic case of a logical text is a mathematical proof. The last two paragraphs of the text "The Pythagorean Theorem" in Lesson 4 (p. 19) are an example of a mathematical proof. (It's also visual; there are drawings.) The premises are

> Draw any two identical squares. Inside the first square, draw a smaller square whose four corners touch the outside square, forming four identical right triangles. Inside the second square, draw four right triangles identical to the four right triangles in the first square, only this time tuck them into the corners of the square as shown in the drawing below.

The conclusions the author draws are

> That leaves two smaller squares inside the second square.
> The smaller square inside the first square is a square on the hypotenuse of each of the four right triangles. The smaller squares inside the second square are squares on the other two sides of the right triangles. Since the bigger squares are identical and the right triangles are identical, the area of the smaller square inside the first square must equal the sum of the areas of the two smaller squares inside the second square.

The text says, in effect, "*If* you make the drawing this way, *then* all this must be true."

You'll find logical texts in many areas of life besides mathematics: science, politics, business, sports—wherever writers try to persuade readers to agree with them.

Often the best graphic organizer for a logical text is a semantic web or a flowchart or a combination of both. The arrows in the flowchart of a logically organized text mean that something follows something else logically, not necessarily in time.

Some special cases of logical organization are concept and definition and concept on concept. **Concept and definition** is a common logical pattern of organization for text. One way to define a concept is to place the concept to be

Structured Notes *(continued)*

defined in a class and then say how it differs from the other members of that class. For example, a bicycle is a vehicle that has two wheels and is powered by pedals. In other words, a bicycle belongs to the class of all vehicles. It is different from other vehicles like automobiles and motorcycles because it is pedal powered and has two wheels. If a concept is defined this way, you probably should copy the definition word-for-word into your notes.

Other ways to define a concept are by listing examples (bicycles are those things that Betty, Juan, and Malik ride to school) or by showing a picture of it (dictionaries often do this).

One way to graphically organize a complex definition is with a Venn diagram like this.

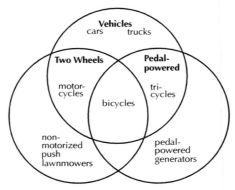

For a simpler definition, an outline or a semantic web may be satisfactory.

Concept on concept is a logical pattern where a writer defines a concept, then defines a second concept in terms of the first and a third concept in terms of the second: bicycles, racing bicycles, triathlon racing bicycles. A flowchart may be the best graphic organizer for a concept-on-concept text.

Structured Notes in Action

Here's a logical text. We'll use it to demonstrate the structured notetaking reading strategy.

> ### How to Win a Shell Game
>
> There is a set of similar games where one player offers the other three choices. If you pick the right one, you win. If you pick one of the two wrong ones, you lose. In the playing-card version, sometimes called three-card monte, three aces are dealt facedown, one black and two red. If you correctly pick out the black ace, you win; if you don't, you lose. Another version is played with three nutshells and a pea. You win if you pick the shell that has the pea under it. The television game show called *Let's Make a Deal* is
>
> *(continued)*

Structured Notes (continued)

How to Win a Shell Game (continued)

another version of the same game. The contestant is asked to choose one of three doors and gets to keep whatever is behind the chosen door. Behind one of these doors is a car; behind the other two are goats.

Assuming they are not rigged—an assumption you should never make if you are asked to play for money with a stranger—these games present an interesting probability problem.

Suppose you are playing the shell version. You pick one shell. Your opponent,—who, of course, knows where the pea is—turns over one of the shells that does *not* have the pea under it, then asks if you would like to stick with your first choice or switch to the other one. Should you switch?

One way to think about this problem is to consider the sample space, which your opponent has altered by turning over one of the losing shells. In doing so, he or she effectively removes one of the two losing shells from the sample space.

If there are three shells, one with a pea under it and two without peas—let's call them A and B—then there are three possible scenarios.

1. You choose the shell with the pea under it. Your opponent turns over either shell A or shell B. If you switch, you lose. If you stick with your original choice, you win.
2. You choose shell A. Your opponent turns over shell B. If you switch, you win. If you stick with your original choice, you lose.
3. You choose shell B. Your opponent turns over shell A. If you switch, you win. If you stick with your original choice, you lose.

Each of these three scenarios has a $\frac{1}{3}$ chance of occurring because you are equally likely to begin by choosing any one of the three shells. In two of them, you win by switching. In one of them, you lose by switching. Thus, the probability of winning is $\frac{2}{3}$ if you switch, which means that you should *always* switch.

This result of $\frac{2}{3}$ may seem counterintuitive. We may believe that the probability of winning should be $\frac{1}{2}$ once we have been shown that the pea is not under shell A (or shell B). Many people reason that since there are two shells left, one of which must conceal the pea, the probability of winning must be $\frac{1}{2}$. This would mean that switching shells would not make a difference. As we've shown through the three different scenarios, however, this is not the case.

One way to convince yourself that $\frac{2}{3}$ is the correct probability is to play the game with a friend. Keep track of how often you win by switching shells (or cards) and by not switching.

Structured Notes (continued)

One way to take notes on this text—not the only way—would be to make a sketch of the game and then make a table to represent the three possible outcomes if you switch shells. Like this.

How to Win a Shell Game

You choose a shell. Opponent turns over shell A or B, showing no pea. Invites you to switch. Should you switch? Three possibilities are

	You choose	If you switch after seeing A or B, you
1	shell w/pea	Lose
2	A	Win
3	B	Win

You win by switching 2 out of 3 times. You should always switch.

Notice that in making the graphic organizer above, we didn't follow any of the formats previously suggested in this program (semantic web, outlining, flowchart, and so forth). The point is to make the graphic organizer fit the text and not the other way around.

Application Make structured notes to help you understand and retain the information in this selection. During your previewing step, pay particular attention to the organization of the selection. This will help you make your structured notes graphic organizer. When you have finished reading, complete the quiz.

Scientific Notation

Many scientists, especially astronomers, have to deal with huge numbers. The distances between stars and galaxies are so great that astronomers measure them in light years. A light year is the distance light travels in one year—almost ten trillion kilometers (six trillion miles). A trillion, when written out, is 1,000,000,000,000. Ten trillion is 10,000,000,000,000.

To save time and space, such large numbers may be written using a shorter method. The shorthand method is based on the fact that:

$100 = 10 \times 10$
$1,000 = 10 \times 100 = 10 \times 10 \times 10$
$10,000 = 10 \times 1,000 = 10 \times 10 \times 10 \times 10$

The number 100 (10×10) can also be written 10^2. The little elevated 2 to the right of the 10 is called an **exponent** or a **power**. It indicates how many times the 10 is multiplied by itself. Thus, such numbers as 10 (10×1), 100 (10×10), 1,000 ($10 \times 10 \times 10$), 10,000 ($10 \times 10 \times 10 \times 10$), 100,000 ($10 \times 10 \times 10 \times 10 \times 10$), 1,000,000 ($10 \times 10 \times 10 \times 10 \times 10 \times 10$) . . . may be written as $10^1, 10^2, 10^3, 10^4, 10^5, 10^6, \ldots$; that is, as powers of ten. The exponent or power indicates

(continued)

Structured Notes (continued)

Scientific Notation **(continued)**
the number of zeros that follow one, or the number of times that 10 is multiplied by itself.

If we multiply numbers written as powers of ten, we can simply add the exponents to find the product. For example, $10^2 \times 10^2 = 10^4$ because $(10 \times 10) \times (10 \times 10) = 10,000 = 10^4$. Similarly, $10^3 \times 10^4 = 10^7$ because $(10 \times 10 \times 10) \times (10 \times 10 \times 10 \times 10) = 10,000,000 = 10^7$.

If we divide numbers written as powers of ten, we can find the dividend by simply subtracting the exponent in the denominator from the exponent in the numerator.

$$\frac{10^5}{10^2} = 10^{5-2} =$$

$$\frac{10 \times 10 \times 10 \times \cancel{10} \times \cancel{10}}{\cancel{10} \times \cancel{10}} =$$

$$10 \times 10 \times 10 = 10^3$$

Two of the tens in the denominator cancel two of the tens in the numerator because $10 \div 10 = 1$. Multiplying or dividing a number by 1 does not change its value.

The number 10 may be written as 10 or as 10^1. Any number raised to the first power is the number itself. For example, $2^1 = 2$, $5^1 = 5$, $9^1 = 9$, $10^1 = 10$.

Any number raised to the zero power is 1. Therefore, $2^0 = 1$, $5^0 = 1$, $8^0 = 1$, $10^0 = 1$. We can show this with an equation:

$$10^4 \div 10^4 = \frac{10 \times 10 \times 10 \times 10}{10 \times 10 \times 10 \times 10} = 10^0 = 1$$

Any number can be written as the product of a number between 1 and 10 (called the **coefficient**) and a power of ten. For example, because 150 is 1.5×100, it can be written as 1.5×10^2. In the same way, 3,200 can be written as 3.2×10^3, 32,000 as 3.2×10^4, and 560,000,000 as 5.6×10^8, and so on. Numbers expressed in this manner are said to be written in **scientific notation.**

The product of two numbers written in scientific notation can be found by multiplying their coefficients and then their powers of ten. For example:
$$3.0 \times 10^6 \times 4.0 \times 10^5 = 12 \times 10^{11}$$
$$= 1.2 \times 10^{12}$$

To divide numbers written in scientific notation, find the quotient of the coefficients and then the quotient of their powers of ten. For example:
$$2.6 \times 10^4 \text{ m} \div 1.1 \times 10^2 \text{ m} = 2.4 \times 10^2$$

Scientific notation isn't just a shorthand method for writing big numbers. It is also a good way to indicate significant figures. When you see a measurement such as 1,200 meters, you cannot be sure how many significant figures it contains. Was the measurement made to ±1 m, ±10 m, or ±100 m? You have no way of knowing. However, if the number is written as 1.200×10^3 m, you know the measurement contains 4 significant figures. The measurement was made to the nearest meter (± 1 m). If written as 1.2×10^3 m, you know there are only 2 significant figures. The measurement is only good to the nearest 100 m (± 100 m). It could be as high as 1.3×10^3 m or as low as 1.1×10^3 m.

Robert Gardner and Edward A. Shore, *Middle School Math You Really Need*, J. Weston Walch, Publisher, 1997.

Structured Notes (continued)

QUIZ: Scientific Notation

Circle the letter of the best answer.

1. $10 \times 10 \times 10 \times 10 \times 10 \times 10 =$
   - (a) 10^6
   - (b) 1,000,000
   - (c) both of the above
   - (d) neither of the above

2. $10^9 \times 10^3 =$
   - (a) 10^{12}
   - (b) 1,000,000,000,000
   - (c) both of the above
   - (d) neither of the above

3. $3.3 \times 10^6 \times 2.0 \times 10^9 =$
   - (a) 6.6×10^{15}
   - (b) 6,600,000,000,000,000
   - (c) both of the above
   - (d) neither of the above

4. $\dfrac{(9.6 \times 10^4)}{(2.0 \times 10^2)} =$
   - (a) 4.8×10^2
   - (b) 480
   - (c) both of the above
   - (d) neither of the above

5. Should scientific notation be used to write the results of all measurements? When is it not necessary or desirable to use scientific notation?

PART 4
Postreading

LESSON 14
Summarizing and Paraphrasing

Postreading (After Reading)

It may seem as if the strategies in this book are meant to make reading more difficult and time-consuming. They are not. They are meant to make your reading more efficient. If you read something for school or work and a few minutes after you put it down you remember nothing about it, then you have not read efficiently. You may have enjoyed the experience, but you have not accomplished anything. You're going to have to read it again. That's not efficient; it's a waste of time. Once you have practiced the reading strategies in this book enough so that they come naturally to you, you'll find that you get more work done faster by using them than by not using them.

The point of all the strategies in this book is to help you remember what you have read. One way to test how much you remember is to have someone ask you questions about it. That's why we have questions at the end of most of the texts in this program. Another way to test how much you remember is to summarize or paraphrase a text.

Summarizing versus Paraphrasing

A **summary** or **paraphrase** of a text is a much shorter written or spoken version of it that repeats its main points. Summarizing and paraphrasing are both postreading strategies: They are done after reading. The difference between a summary and a paraphrase is that a summary uses some of the same words as the original, sometimes repeating entire sentences from the text being summarized. To paraphrase a text is to retell it in your own words.

In school, you may be asked to summarize or paraphrase a text orally in class, in writing on a test, or in writing on a paper or project. But even if you are never asked to do any of those things, the best way to see how much you have retained of a text is to try summarizing or paraphrasing it to yourself. If you can summarize or paraphrase a text and then see, by looking back at the text or your notes, that you have done so accurately, you know you have learned something.

Summarizing and Paraphrasing (continued)

Summarizing and Paraphrasing in Action

Buildings Built to Waste Energy

Big buildings use energy to control the temperature and humidity inside, to keep them cool in the summer and warm in the winter. They lose that energy through their containing surfaces. So if we want to build an energy-efficient building, one that stays comfortable with the least energy cost, what shape should it be?

The shape that contains the most volume with the least surface—that is, has the least possible surface-to-volume ratio—is a sphere. If you double the diameter of a spherical structure, you increase its contained atmosphere eightfold and its enclosing surface only fourfold. When you double the diameter of a sphere, the size of the contained molecules of atmosphere does not change. Therefore, every time you double a spherical structure's diameter, you halve the amount of enclosing surface through which an interior molecule of atmosphere can gain or lose energy. So the most energy efficient shape for a building is spherical, and the bigger the better.

Skyscrapers are built to put the greatest possible volume on small, usually rectangular, parcels of expensive city real estate. The typical result is a tall, flat slab or a rectangular-sectioned tube. Flat slabs have a high surface-to-volume ratio. That's why flat slab fins make good air-cooling motorcycle and light-airplane engines. Tubes have the highest surface-to-volume ratios. Triangular- or rectangular-sectioned tubes have higher surface-to-volume ratios than round-sectioned tubes. Tall slab buildings and rectangular-sectioned, tubular-tower skyscrapers may make efficient use of land, but they waste energy.

Adapted from R. Buckminster Fuller, *Critical Path*. New York: St. Martins Press, 1981.

Summary

Here's our summary. Notice that the last sentence of the original text, which is the conclusion of the author's analysis, becomes the first part of the first sentence of the summary.

> Tall slab buildings and rectangular-sectioned, tubular-tower skyscrapers may make efficient use of land, but they waste energy because their surface-to-volume ratio is very high. The most energy-efficient shape for a building is spherical, and the bigger the better.

Paraphrase

Here's a paraphrase. Again, we begin with the main point of the text, but this time we don't use the same words.

> Skyscrapers are badly designed from the point of view of energy efficiency. If you want a building to stay warm in the winter or cool in the summer, the best design is a sphere. The surface of a building is where warmth or coolness is lost. A sphere holds the most volume with the least surface area. The tall, thin, tubular shape of skyscrapers has much more surface for the same volume and therefore loses much more energy.

 © 2002 J. Weston Walch, Publisher

Summarizing and Paraphrasing (continued)

Application Try summarizing and then paraphrasing this text.

Simple and Compound Interest

Banks and various companies will pay you to allow them to use your money for their own purposes. The money they agree to pay you or add to the money you invest is called interest. There are two kinds of interest: simple and compound.

To find simple interest all you have to do is multiply the principal by the rate ($p \times r$). The principal is the amount of money with which you start your investment. For example, a person might have $5,000 to invest. The $5,000 is the principal. He or she finds a company that will guarantee an interest of 8% per year. The rate—the simple interest paid per year—is 8%. At the end of one year, the principal will be worth $5,000 + the interest. The simple interest on $5,000 after one year is:

$$\$5,000 \times .08 = \$400$$

(Notice that 8% has been changed to .08 because 8% means 8/100 or 0.08.)

At the end of one year your money has grown to $5,000 plus the interest of $400, giving a total of $5,400.

At the end of the second year, the principal once again earns 8%, or $400. If you don't spend any of the money, you will now have a total of $5400 + $400 = $5800.

Compound interest is similar but better for the investor. With compound interest, the interest is added to the principal each year (or possibly more often). For example, if $5,000 were invested at 8% compounded annually, then at the end of one year the principal would be worth $5,400. The 8% interest for the second year would be paid on $5,400, not $5,000. At the end of the second year, the principal would have increased to:

$$\$5,400 + (\$5,400 \times 0.08) = \\ \$5,400 + \$432 = \$5,832.$$

As you can see, you are $32 richer with compound interest than with simple interest.

Interest may be compounded more often than once a year. Suppose the company agrees to pay 8% interest compounded semi-annually (every 6 months). At the end of 6 months, the principal will have grown to:

$$\$5,000 + (\$5,000 \times 0.08/2) = \$5,200.$$

At the end of the first year, the principal will be worth:

$$\$5,200 + (\$5200 \times .08/2) = \\ \$5,200 + \$208 = \$5,408.$$

Compounding the interest on an investment makes a big difference when it is done over a long period of time!

Robert Gardner and Edward A. Shore, *Middle School Math You Really Need*, J. Weston Walch, Publisher, 1997.

Summarizing and Paraphrasing (continued)

Summary

Paraphrase

PART 5
Reading in Mathematics

Lesson 15
Common Patterns and Features of Mathematics Writing

Ode to Math Writing

When you need to communicate with other people about arranging things in order, counting things, measuring things, the shape of things, and come to agreement about things in the observable, physical world that we all share, mathematics is what you use.

- Math is clear. Mathematical terms are precisely defined.
- Math is general. The number 4 can stand for 4 apples, 4 seconds, 4 dinosaurs, 4 dollars, 4 asteroids, 4 gigabytes—4 of anything.
- Math is concise. $a^2+b^2=c^2$ says a great deal in eight characters. To say the same thing in English could take a hundred characters or more and wouldn't be as clear.
- Math is flexible. Math offers many ways to say the same thing.
- Math is logical. If you understand math, it's much easier to follow the logic of an argument written in math than the same argument written in English.

All these wonderful qualities that make the language of mathematics so useful can also make math hard to read. Because math is so concise, you must read math more slowly than you read non-mathematical English. Every word, number, and symbol counts. You must read math over and over until you understand it. Context clues are not always available in math writing. Mathematical terms are precisely defined. If you don't know exactly what a math word means, look it up.

Common Patterns and Features

Math writing may be organized in many ways. Below is a list of some of the most common. Recognizing the organizational pattern of a reading can help you choose the best reading strategy.

- Main idea and details
- Visual texts
- Concept and definition
- Concept on concept
- Steps in a process

LESSON 16
Main Idea and Details

Main Idea and Details

Many texts are organized on the pattern of main idea and details. The **main idea** is the point of the text. The **details** are there to support the point, to explain it, to clarify it, to give evidence for it, to give examples of it, and so on. Main ideas tend to be more abstract or general; details tend to be more concrete and specific. One way to tell the difference between a main idea and a detail is to ask yourself, "What if this sentence appeared on the page all by itself?" If the answer is, "That might be interesting. I wonder what it means," or "I wonder if it's really true," then it's probably a main idea. If the answer is, "So what?" then it's probably a detail.

No one can possibly remember everything they read. When you read, you need to make intelligent choices about what to focus your attention on, take notes on, and remember, and what to read and forget. In making your choices, you need to keep two purposes in mind: your purpose and the author's. The main idea of a text reflects the author's purpose in writing, the point he or she wants to get across. Usually, the main idea is what you want to remember—but not always. Sometimes, your purpose is to get certain details and forget the author's main idea. Follow your own purpose.

However, assuming that your purpose in reading is more or less aligned with the author's purpose in writing, the most appropriate graphic organizer for a main-idea-and-details text is usually an outline or a semantic web. Both require you to identify the main idea. But you may find another graphic organizer more useful to you.

Application

The following text is organized by main ideas and details. Read the article, then answer the questions.

> ### Leap Years
>
> If you live on the earth, there are two units of time that come naturally: the day and the year. A day is the time it takes for the earth to rotate on its axis. A year is the time it takes for the earth to revolve around the sun. But a solar year is not a whole number of days; it is 365.242195 days. That's a problem for people who make calendars.
>
> To make a calendar that keeps the
>
> *(continued)*

 © 2002 J. Weston Walch, Publisher

Content-Area Reading Strategies: Mathematics

Main Idea and Details *(continued)*

Leap Years *(continued)*

seasons from slowly drifting forward or backward over time, the astronomer Sosigenes, who worked for the Roman Emperor Julius Caesar, invented the leap year in 46 B.C.E. The Julian calendar, as it came to be called, saved up the extra decimal fraction of a day, 0.2422, for four years, then added one extra day (February 29) to every year that is a multiple of four. That helped, but it did not completely solve the problem. An extra day every four years, 0.25 days per year, is too much; $4 \times 0.2422 = 0.9688$ day, 0.0312 days or about 45 minutes less than a whole day.

Three quarters of an hour in four years may not seem like much, but by 1582, the calendar year had slipped backward a full 12 days from the year measured by the sun. Pope Gregory, the head of the Roman Catholic Church, one of the few functioning international organizations of the time, consulted with his astronomers. He decreed that ten days should be added back to the calendar immediately. He further decreed that years that are multiples of 100—1700, 1800, 1900—would henceforth *not* be leap years unless they are also multiples of 400. Following this rule, 1600 and 2000 were leap years.

Does the Gregorian calendar, as it is called, solve the problem of making a whole-day calendar out of a fractional-day year? Not entirely. There is still an error of 0.12 days every 400 years. It has been suggested that our descendants could reduce the error further by making the year 4000 and all multiples of 4000 *not* leap years.

Adapted from Donald H. Menzel, *A Field Guide to the Stars and Planets.* Boston: Houghton Mifflin Company, 1964.

1. What is the main idea of the first paragraph?

2. What is the main idea of the second paragraph?

3. What is the main idea of the third paragraph?

4. What is the main idea of the fourth paragraph?

5. What is the main idea of the whole article?

LESSON 17
Visual Texts

Some texts include visual information—maps, photographs, illustrations, graphs, tables, diagrams, geometric figures, or the like. Sometimes the visual information supports the text; sometimes the text supports the visual information. Whichever is the case, when you preread, go to the visual information first, then to whatever text is closely tied to the visuals—titles, captions, legends, callouts. Usually you can get a sense of what the text is about more quickly that way than by reading the text first.

If the visual information is important—and it usually is—get it into your notes as best you can, ideally by photocopying it, otherwise by sketching it.

Youth Crime Drop

According to the most recent FBI statistics, after rising for a decade, the rate of violent crime in the United States suddenly began to go down in the mid 1990s. While criminal behavior has always been more prevalent among young people than adults, just as young people were responsible for the previous increase in violent crime, young people are also responsible for the recent decline.

Number of Arrests

The FBI's data on youth crime is actually arrest data. It would be better if data on crimes reported to the police were available sorted by age, but if no

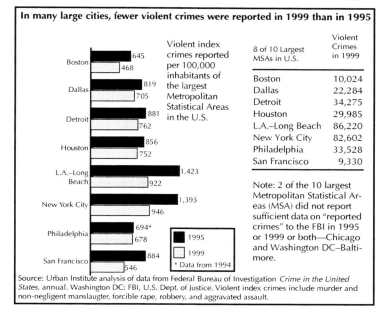

(continued)

© 2002 J. Weston Walch, Publisher   **67**   Content-Area Reading Strategies: Mathematics

Youth Crime Drop (continued)

one is arrested for a crime, the age of the perpetrator is unknown. So arrest data is the best data we have. It tells us something about the contribution of young people to the crime problem as compared to adults.

In 1999, U.S. law enforcement agencies made about 14 million arrests. Of these, 17 percent involved juveniles under age 18, and 28 percent involved youths between the ages of 18 and 24. Arrests of young people for many of the most serious offenses fell substantially between 1995 and 1999. During this period arrests of juveniles for murder dropped 56 percent; robbery, 39 percent; burglary, 23 percent; and motor vehicle theft, 35 percent.

The total number of arrests involving juveniles in 1999, 2.5 million, was only 9 percent lower than in 1995. However, this is because the decrease in more serious crimes was offset by increases in juvenile arrests for less serious offenses like driving under the influence (up 36 percent), alcoholic beverage law violations (up 31 percent), and arrests for curfew violations (up 9 percent).

Arrest Rates

Is this decline in youth arrests for serious crimes a result of demographic changes? No. Controlling for changes in population, the rate of decline in juvenile arrests was steeper that in other age groups. The violent crime (murder, forcible rape, aggravated assault, and robbery) arrest rate fell among all age groups between 1995 and 1999. The rate of violent crime arrests for adults was relatively unchanged throughout the 1990s, but the juvenile arrest rate in 1999 was about two-thirds the rate of 1995.

The experts are not entirely sure why youth crime decreased during the late 1990s, but various explanations have been offered: a strong economy, changing demographics, changes in the market for illegal drugs, changes in the availability of firearms, tougher prison sentences, innovations in policing, and a growing intolerance for violent behavior on the part of society.

Juvenile arrests declined 9% between 1995 and 1999, with larger decreases in violent offenses.

	National estimate of juvenile arrests, 1999	% Change: 1995–99
All offenses	2,468,800	–9
Violent Crime Index offenses	**103,900**	**–23**
Murder/non-negligent manslaughter	1,400	–56
Forcible rape	5,000	–11
Robbery	28,000	–39
Aggravated assault	69,600	–13
Index property	**541,500**	**–24**
Burglary	101,000	–23
Larceny–theft	380,500	–23
Motor vehicle theft	50,800	–35
Arson	9,200	–19
Selected other offenses		
Other assaults	237,300	2
Weapons	42,500	–27
Drug abuse violations	198,400	1
Driving under the influence	23,000	36
Liquor laws	165,700	31
Disorderly conduct	176,200	–3
Curfew/loitering	170,000	9
Runaways	150,000	–28

Source: Urban Institute analysis of data from Federal Bureau of Investigation. *Crime in the United States,* annual. Washington DC: FBI, U.S. Department of Justice.

(continued)

Visual Texts (continued)

Youth Crime Drop (continued)

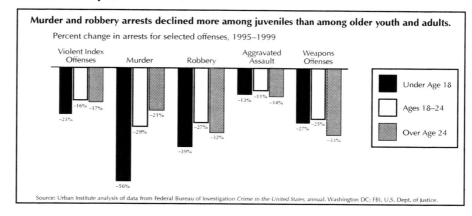

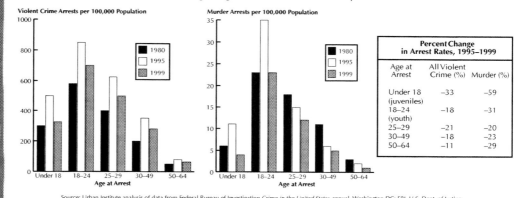

Adapted from Jeffery A. Butts. Washington: Urban Institute, 2000. (www.urban.org)

Sometimes authors are so eager to make a point that they exaggerate what the data in their graphs and tables actually says. Sometimes, for the same reason, they ignore interesting facts in their data. Does this author do either of those things? Check each statement the author makes in the text against the data in the table and the graphs to see if it is a fair representation of the data. Examine the table and the graphs to see if there is any interesting information there that the author ignored.

Did any of the information in the text or the visuals surprise you? Explain.

Lesson 18
Concept and Definition

Concept and Definition

Concept-and-definition texts are common in mathematics. As you know, one advantage of math as a language over natural languages like English is that the vocabulary of math is so clearly defined that there is seldom any dispute over the meaning of a mathematical expression. This doesn't happen naturally. Mathematicians take pains to get definitions exactly right.

Application

Here's an example of a mathematical concept and definition. Read the text. Then answer the questions that follow.

> ### Converse
>
> The relation of wife to husband is called the *converse* of the relation of husband to wife. Similarly *less* is the converse of *greater*, *later* is the converse of *earlier*, and so on.
>
> Generally, the converse of a given relation is that relation which holds between y and x whenever the given relation holds between x and y.
>
> Bertrand Russell, *Introduction to Mathematical Philosophy.* London: George Allen & Unwin, Ltd., 1919, p. 16.

Probably the best graphic organizer for this text would be a semantic web with the definition itself (the last sentence) at the center and arrows pointing to examples.

1. Is the concept clear to you from this definition?

2. Are all the terms used in it clear?

3. Can you think of more examples of converse relations?

LESSON 19
Concept on Concept

Concept on Concept

Concept-on-concept texts are common in mathematics. Mathematics, as well as being an art form, a science, a set of tools, and a language, is a game. Like all games, it has rules. The rules of mathematics, like the rules of any game, must be both exhaustive and consistent. *Exhaustive* means that the rules cover every situation that can come up in the game; *consistent* means that the rules never contradict each other.

To ensure that the rules are consistent, they must be written using terms that are defined consistently. To ensure that the definitions of terms are consistent, it helps to start with a very short list of terms and use them to define all other terms. That's what mathematicians do.

Here's a text that defines several terms, each definition building logically on the last. Probably the best graphic organizer for a text of this kind is a series of semantic webs connected like a flowchart.

Application

Read the text. Then answer the questions that follow.

Arithmetic Sequences

A child first becomes aware of numbers through counting. Arranged in order, the counting numbers form a number sequence.

1 2 3 4 5 6 7 8 9 10 . . .

A *number sequence* is an arrangement of numbers in which each successive number follows the last according to a uniform rule. The numbers are the *terms* of the sequence. For the sequence of counting numbers, the rule is, "add one to each term to get the next term." In symbols, if n represents any term of the sequence, the next term is $n + 1$. The fact that the sequence of counting numbers can be continued indefinitely is indicated by the three dots following the last term shown. A mathematician once said, "A number sequence is like a bus; nobody ever doubts that there is always room for one more."

When a parachutist jumps from an airplane, the distances in feet traveled during each of the first few seconds are

16 48 80 112 144 . . .

This sequence is similar to the sequence of counting numbers in that each successive term can be found by adding the same number to the preceding term.

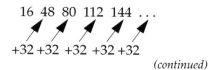

(continued)

Concept on Concept (continued)

> **Arithmetic Sequences (continued)**
>
> A number sequence in which each successive term may be found by adding the same number is an *arithmetic sequence*.
>
> If the number added is positive, the sequence grows at a constant rate. Here is an arithmetic sequence formed by adding 3.
>
> 2 5 8 11 14 17 20 23 ...
>
> If the number added is negative, the sequence shrinks at a constant rate. This arithmetic sequence is formed by adding –2:
>
> 11 9 7 5 3 1 –1 –3 ...
>
> The number added to each term of an arithmetic sequence to form the next term is also the difference between each pair of successive terms. It is called the *common difference* of the sequence. The common difference of the first sequence is 3, and the common difference of the second sequence is –2.

1. What is the order of concepts?

2. Is each concept clearly explained before moving on?

3. Is the relationship between one concept and the next clear?

4. Was this a good choice of organization?

LESSON 20
Steps in a Process

Steps in a Process

Math is full of step-by-step processes that lead to certain results. Thus, mathematicians often refer to them using a word that, like much of our mathematics, comes from Arabic: *algorithm*. An algorithm is a step-by-step method of calculating a certain result.

You already know several algorithms. For example, when you do long division, you are following a step-by-step procedure that starts with a number to be divided by another number and leads to a quotient.

Many algorithms exist in science, engineering, accounting, statistics, and all the various disciplines that use mathematics. Today, many algorithms are programmed into computer software so that you can get the result without doing the work. Spreadsheet software enables you to set up your own algorithm and have a computer follow it.

Application

When you read a step-by-step process text, the best graphic organizer for your notes is usually a type of structured notes: a flowchart. Create one as you read this text. Then answer the question that follows the text.

Operating Costs of Electrical Appliances

A century ago, food was kept in an icebox—an insulated box that held a large piece of ice. Food was cooked on a wood- or coal-burning stove made of cast iron. Clothes were washed on a corrugated board, known as a scrub board, and dried on a line in a sunny area or stretched across one end of the kitchen. Dishes were washed in a basin in the sink using water pumped by hand from a well and floors were swept, not vacuumed, with a broom.

Today, life is much easier physically. We have electrical appliances—refrigerators, stoves, washers, dryers, dishwashers, vacuum cleaners, and more. But these appliances come at a price.

Not only do we have to buy the appliances, we have to pay for the electrical energy to operate them.

Fortunately, you can make a reasonable estimate of the annual cost to operate any appliance. To do so you need to know three things: (1) The approximate length of time you expect the appliance to be used each year; (2) The appliance's wattage rating (the electrical energy it uses per hour); (3) The cost of the electrical energy supplied by your local power plant in cents per kilowatt-hour.

Suppose you want to know how much it will cost to operate a TV set for a year. Somewhere on the TV set

(continued)

Steps in a Process (continued)

Operating Costs of Electrical Appliances (continued)

you will find its operating wattage. It might be 200 watts (W), which equals 0.2 kilowatts (kW) because one kilowatt is equal to 1000W. This is how much energy it takes to power the TV set for one hour.

You need to estimate how many hours per year the set is turned on. The best way, of course, would be to put a meter on the power cord. Or you could keep a TV diary for a week and multiply the week's total hours by 52. If you can't do that, you can make a rough estimate based on your knowledge of TV watching in your household, or use the estimated national average of 4 hours per day and multiply by 365.

At this writing, Massachusetts Electric charges its residential customers $0.11 per kilowatt-hour. To find out what your electric company charges, look at your latest bill.

To calculate the kilowatt-hours of energy needed to operate your TV set for a year, multiply the kilowatts needed to operate the TV by the number of hours you estimate it will be used in a year.

To estimate the operating cost, multiply the kilowatt-hours of energy needed per year by the cost of the energy per kWh.

Adapted from Robert Gardner and Edward A. Shore, *Middle School Math You Really Need*, J. Weston Walch, Publisher, 1997.

Can you express this algorithm as a formula? Be sure to define every term in your formula clearly. Each definition should specify the units by which it is measured.

Lesson 21
Review

In this book, you have learned some strategies for reading nonfiction more efficiently. Here's a quick review of the strategies you have learned and practiced.

Prereading **The 4 Ps**

- **Preview.** Scan the text. Find out as much as you can about what you are going to read without actually reading it.
- **Predict.** Answer this question: Based on what you saw during the previewing stage, what do you think the reading will be about?
- **Prior knowledge.** What do you already know about this topic?
- **Purpose.** What do you want to get from this reading?

Reading **The Five Reading Strategies**

- **KWL.** What I KNOW, What I WANT to Know, What I LEARNED. Take notes in a three-column chart.
- **SQ3R.** Survey, Question, Read, Recall, Reflect. The survey and question columns are similar to the 4 Ps.
- **Semantic web.** During prereading and reading, draw a diagram using circles and lines to connect ideas.
- **Outline.** Note information using indention to show importance of ideas.
- **Structured notes.** Create your own graphic organizer based on the structure of the reading selection.

Postreading **Summarizing and Paraphrasing**

Both strategies ask you to think about what you have read. Writing this helps solidify your new knowledge.

- **Summarize.** Write a condensed version of the key points, using the words from the text.
- **Paraphrase.** Retell the main points in your own words.

 © 2002 J. Weston Walch, Publisher

PART 6
Practice Readings

Reading A

Use a graphic organizer to record important information in the selection below. Read the selection carefully, making notes as you read. When you have finished, be sure to write a brief summary or paraphrase of the material. Then check your understanding by taking the quiz that follows the reading.

Supply and Demand

Everyone has observed that the higher the price charged for an article, the less of it will be sold. And the lower its price, the more units people will buy. Thus there exists at any one time a definite relation between the price of a good such as wheat and the quantity demanded of that good. The following hypothetical table relating *price* and *quantity demanded* is an example of what economists call a "demand schedule." At any price, such as $5 per bushel, there is a definite quantity of wheat that will be bought by all the consumers in the market—in this case 9 (million) bushels per month. At a lower price, such as $4, the quantity bought is even greater, being 10 (million) units. From [the preceding table] we can determine the *quantity bought at any price*, by comparing Column 1 with Column 2.

The total number of dollars of revenue received for the sale of wheat is equal to the number of bushels sold times the price per bushel. Thus if the price is $5 and the quantity sold 9 (million) bushels, then the total revenue received from the sale of wheat is 45 million dollars. . . .

The Demand Curve

The same numerical data [in the table preceding] can be given a more graphic interpretation. On the vertical scale [in the graph] we represent the various prices of wheat, measured in dollars per bushel. On the horizontal scale, we measure the quantity

	1	2	3
	Price of wheat per bu. P (1)	Quantity demanded, million bu. per month Q (2)	Value of sales, millions of dollars per month, price × quantity (3) = (1) × (2)
A	$5	9	$45
B	$4	10	$40
C	$3	12	$36
D	$2	15	$30
E	$1	20	$20

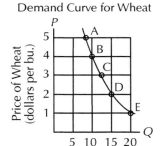

Demand Curve for Wheat

(continued)

Reading A (continued)

Supply and Demand (continued)

of wheat (in terms of bushels) that will be bought.

Just as a city corner is located as soon as you know its street and avenue, so is a ship's position located as soon as you know its latitude and longitude. Similarly, to plot a point on this [graph], we must have two coordinate numbers, a price and a quantity. For our first point A, corresponding to $5 and 9 million bushels, we move upward 5 units and then over to the right 9 units. A circle marks the spot A. To get to the next circle, at B, we go up only 4 units and over to the right 10 units. The last circle is shown by E. Through the circles we draw a smooth curve . . .

This [graphic representation] of the demand schedule is called the "demand curve." Note how quantity and price are inversely related. The curve slopes downward, going from northwest to southeast. This may be given a name: the *law of diminishing demand*, as price rises. This law is true of almost all commodities: wheat, electric razors, cotton, gasoline, and cornflakes. . . .

A number of obvious reasons can be given for the prevalence of the law of diminishing demand: (1) At lower prices the consumer's dollar goes further. [The consumer] can *afford* to buy more. (2) At lower prices, [the consumer] will *want* to buy more. Cheap wheat means that people will want to get their calories by *substituting* bread for potatoes and white bread for rye bread. (3) Secondary uses of wheat such as for grain alcohol in producing synthetic rubber will be possible only at sufficiently low prices. Each reduction in price will tend to bring in some new buyers and some new uses by old buyers, until finally the price is zero and wheat is used with the lavishness of a free good.

> **The law of diminishing demand:** If the price of a commodity is cut, more of it will be demanded. To say the same thing in another way, if a greater quantity of a commodity is thrown on the market, it can be sold only at a lower price.

From *Economics: An Introductory Analysis* by Paul A. Samuelson. ©1951 by McGraw-Hill Book Company, Inc. Reproduced with permission of The McGraw-Hill Companies.

Reading A *(continued)*

QUIZ: Supply and Demand

Circle the letter of the best answer.

1. In this text, the word *schedule* means a table showing the
   (a) amount of wheat sold.
   (b) price of wheat.
   (c) relation between one variable quantity and another.
   (d) timing of events.

2. When the price of wheat goes up,
   (a) consumers buy less wheat.
   (b) farmers sell less wheat.
   (c) Both of the above.
   (d) Neither of the above.

3. Price and quantity are inversely related. If they were not,
   (a) the demand curve would go from southwest to northeast.
   (b) as price increased, quantity would also increase.
   (c) Both of the above.
   (d) Neither of the above.

4. Secondary uses of wheat, such as for grain alcohol in producing synthetic rubber, will be possible only at sufficiently low prices because if the price of wheat were higher,
   (a) the makers of synthetic rubber would buy something else to make grain alcohol with.
   (b) synthetic rubber produced using grain alcohol made with wheat would cost too much.
   (c) Both of the above.
   (d) Neither of the above.

5. If the wheat crop were largely wiped out by pests, what would happen to the amount of wheat sold and the price of wheat?

Reading B

Use a graphic organizer to record important information in the selection below. Read the selection carefully, making notes as you read. When you have finished, be sure to write a brief summary or paraphrase of the material. Then check your understanding by taking the quiz that follows the reading.

The Three Suspects

Caroline Jones is the chief of the police force of a major city. Three men are brought in for questioning. From information received from reliable sources, Caroline knows that one of the three is a sane, law-abiding citizen, one is a harmless eccentric, and one is a dangerous criminal. Her sources cannot tell her which is which.

Furthermore, she knows that the sane, law-abiding citizen always tells the truth, the harmless eccentric never tells the truth, and the dangerous criminal sometimes tells the truth and sometimes doesn't.

The detectives interrogate all three suspects and report to Chief Jones as follows.

Mr. One says, "Mr. Two is a sane, law-abiding citizen. You should let him go."

Mr. Two says, "I am a dangerous criminal. If you let me go, I will kill thousands."

Mr. Three says, "Mr. Two is a harmless eccentric. You should let him go."

Chief Jones thinks a moment and says, "Lock up Mr. One. Take Mr. Two to the Mental Health Center. Let Mr. Three go with our apologies."

Later, confirming evidence proves that the identifications are correct. The amazed detectives ask Chief Jones how she knew. "Logic," she says.

"If Mr. One were telling the truth when he said that Mr. Two was the sane, law-abiding citizen, then Mr. Two would agree. But Mr. Two didn't agree. Therefore, Mr. One was not telling the truth, and Mr. One is not the sane, law-abiding citizen. Therefore, the sane, law-abiding citizen must be either Mr. Two or Mr. Three.

"If Mr. Two were the sane, law-abiding citizen, then he would say so, because the sane, law-abiding citizen always tells the truth. But Mr. Two didn't say so. Therefore, Mr. Two is not the sane, law-abiding citizen. And therefore, the sane, law-abiding citizen must be Mr. Three.

"Mr. Three said that Mr. Two was the harmless eccentric. And because Mr. Three, the sane, law-abiding citizen, always tells the truth, it must be true.

"If Mr. Three is the sane, law-abiding citizen and Mr. Two is the harmless eccentric, then Mr. One must be the dangerous criminal."

Reading B *(continued)*

QUIZ: The Three Suspects

Circle the letter of the best answer.

1. To *interrogate* means to
   (a) question.
   (b) arrest.
   (c) detain.
   (d) answer.

2. Chief Jones knows that Mr. Three is the sane, law-abiding citizen because
   (a) Mr. Two did not tell the truth.
   (b) Mr. One did not tell the truth.
   (c) Both of the above.
   (d) Neither of the above.

3. Chief Jones knows that Mr. Two is the harmless psychiatric patient because
   (a) Mr. Three said so.
   (b) Mr. Two said so.
   (c) Mr. One said so.
   (d) None of the above.

4. Chief Jones knows that Mr. One is the dangerous criminal because
   (a) Mr. Two is not.
   (b) Mr. Three is not.
   (c) Both of the above.
   (d) Neither of the above.

5. Could a case like this happen in real life? Chief Jones solved the case by reasoning logically from certain premises. How realistic are those premises?

Reading C

Use a graphic organizer to record important information in the selection below. Read the selection carefully, making notes as you read. When you have finished, be sure to write a brief summary or paraphrase of the material. Then check your understanding by taking the quiz that follows the reading.

Sampling Red Blood Cells

Suppose you need to know the number of objects in a container that is too large to allow you to count individually. You can first sample a small measurable volume and count the objects. Then find the volume of the large container and you'll be able to find the total number of objects.

For example, suppose that we want to know the number of beans in a large jar that measures 50 cm across its circular base and stands 150 cm high. We remove a small sample to a 500-cubic-centimeter beaker and count 378 beans in the beaker. That means that there are $2 \times 378 = 756$ beans in every 1,000 cm^3. The volume of the large vessel is the volume of a cylinder of radius 25 cm and height 150 cm:

Volume = $\pi r^2 h = 3.14 \times (25\text{cm})^2 \times 150$ cm = 294,375 cm^3.

Since every 1,000 cm^3 contains 756 beans, there are $294.4 \times 756 \approx 223,000$ beans in the large container.

A drop of blood when spread out on a glass slide and viewed through a microscope reveals a vast number of tiny dumbbell-shaped cells. These cells are red blood cells. They are the cells that carry oxygen from our lungs to the other cells of our bodies. One look through the microscope would convince you that it would be not only tedious but also impossible to count all the red blood cells in your body. When a blood test reports a red blood cell count of so many million per cubic millimeter, it is based on an estimate made by a laboratory technician.

To make such an estimate, a technician can draw a tiny volume of a patient's blood into a small pipette. The blood is diluted by mixing one part blood with 199 parts saline (salt) solution. After the blood and saline are thoroughly mixed, a drop of the diluted blood is placed in a small, shallow counting chamber that has a volume of exactly 0.10 cubic millimeter. The chamber is divided by cross-rulings into 400 equal spaces. The technician then looks at the diluted blood in the chamber through a microscope and counts the number of red blood cells in 80 of the 400 spaces.

❑ Let's assume that the technician counts a total of 450 red blood cells in the 80 spaces. How does he or she then estimate the number of red blood cells per cubic millimeter?

(continued)

© 2002 J. Weston Walch, Publisher

83

Content-Area Reading Strategies: Mathematics

Reading C *(continued)*

Sampling Red Blood Cells (continued)

The technician counted the cells in 80 of the 400 cells, or 80/400 of the cells in the chamber. Since 80/400 = 1/5, only 20 percent of the cells were actually counted. To estimate the total number of red blood cells in the chamber, we would have to multiply the number actually counted by 5. But the volume of the chamber was only 0.10 mm3. To estimate the total number of cells in one cubic millimeter of the diluted blood, the technician would have to multiply by 10, since the chamber was only $1/10$ mm^3. So far, the number of cells counted has been multiplied by 50 (5×10). But remember, the blood was diluted 1:200 when it was mixed with saline solution. Consequently, to make an accurate estimate, the technician would have to multiply again by 200. Altogether then, the number of cells actually counted would have to be multiplied by $5 \times 10 \times 200 = 10{,}000$ to estimate the number of red blood cells in 1.0 mm^3 of blood.

Robert Gardner and Edward A. Shore, *Math You Really Need*, J. Weston Walch, Publisher, 1996.

Reading C (continued)

QUIZ: Sampling Red Blood Cells

Circle the letter of the best answer.

1. *Tedious* means
   (a) boring.
   (b) impossible.
   (c) plentiful.
   (d) sanguine.

2. An *estimate* is a measurement that is
   (a) approximately correct.
   (b) definitely wrong.
   (c) exactly correct.
   (d) probably wrong.

3. The blood sample was diluted by mixing it with
   (a) beans.
   (b) blood.
   (c) oxygen.
   (d) saline solution.

4. The technician estimated 4,500,000 red blood cells per cubic millimeter of blood. This means that the patient's actual blood count is probably
   (a) between 4,000,000 and 5,000,000.
   (b) between 4,400,000 and 4,600,000.
   (c) between 4,490,000 and 4,510,000.
   (d) between 4,499,000 and 4,501,000.

5. The text says, "The blood is diluted by mixing one part blood with 199 parts saline (salt) solution." Later it says, "But remember, the blood was diluted 1:200 when it was mixed with saline solution." Is this a mistake? Does it matter? Why or why not?

Reading D

Use a graphic organizer to record important information in the selection below. Read the selection carefully, making notes as you read. When you have finished, be sure to write a brief summary or paraphrase of the material. Then check your understanding by taking the quiz that follows the reading.

Archimedes

Archimedes, a Greek who lived in Syracuse on the island of Sicily from 287 to 212 B.C.E., was one of the greatest mathematicians of all time. He discovered some of the properties of solid geometric figures that we study today.

Archimedes applied his knowledge of mathematics to the design of machines to do practical work. For example, he invented the pulley, a device still used today to lift heavy weights by applying a small force. He demonstrated the pulley by single-handedly moving a fully loaded ship. Archimedes also designed weapons to defend Syracuse against the invading Roman armies. His catapults hurled huge rocks over the walls of the city.

Archimedes is best known for Archimedes' Law, which says that when you immerse an object of a certain volume in a fluid—say, a submarine in water or a helium balloon in air—the force pushing it upward is equal to the weight of an equal volume of the fluid.

Archimedes discovered this principle when Hiero, the king of Syracuse, asked Archimedes to help catch a dishonest goldsmith. The king had given the goldsmith some gold and asked him to make a crown. The goldsmith had given back a crown that weighed exactly the same as the gold he had been given. However, the king suspected that the goldsmith had kept some of the king's gold and replaced it with silver or some other metal. But how could he prove it?

While taking a bath, Archimedes thought deeply about the question. He noticed that when he got into the tub, water displaced by his body splashed over the side. That was it! Without stopping to get dressed, he ran all the way to the king, shouting "Eureka," which is Greek for "I've got it!"

Archimedes devised the following test. Measure out a quantity of pure gold equal in weight to the lump of gold originally given to the goldsmith. Immerse it in water. Measure how much water is displaced. Immerse the crown in water. Measure how much water is displaced by the crown. Compare the amount of water displaced by the gold to the amount of water displaced by the crown.

Gold is the densest of all metals. It weighs more per unit of volume than any other metal. If the pure gold and

(continued)

Archimedes (continued)

the crown displace the same amount of water, then they have the same volume. If the crown displaces more water than the pure gold—which it did—then it has a greater volume. Therefore, the crown has silver or some other less dense metal mixed into it.

Archimedes died doing a math problem. He was concentrating so hard, he didn't even notice that Roman troops had invaded Syracuse. Marcellus, the Roman general, sent a soldier to get Archimedes because of his expertise. The soldier ordered Archimedes to come along. Archimedes asked the soldier to wait until he finished the proof he was working on. The soldier was so angry he drew his sword and killed Archimedes.

Reading D (continued)

QUIZ: Archimedes

Circle the letter of the best answer.

1. *Immerse* means to
   (a) dunk.
   (b) measure.
   (c) care for.
   (d) purify.

2. Cold air is denser than hot air. Therefore, a cubic centimeter of cold air will weigh _____ a cubic centimeter of hot air.
   (a) more than
   (b) less than
   (c) the same as
   (d) It's impossible to tell.

3. If the crown that the goldsmith made for King Hiero had displaced the same volume of water as the lump of pure gold, we could conclude that
   (a) the goldsmith had cheated the king.
   (b) the goldsmith was honest.
   (c) the gold the king gave the goldsmith was not pure.
   (d) None of the above.

4. If the crown that the goldsmith made for King Hiero had displaced less water than the lump of gold, we could conclude that
   (a) the goldsmith had cheated the king.
   (b) the goldsmith was honest.
   (c) the gold the king gave the goldsmith was not pure.
   (d) None of the above.

5. Describe a time when, like Archimedes, you were so involved in solving a problem or completing a job that you didn't notice what was going on around you and could not bear to stop, even though it made someone else angry.

Blank Graphic Organizers

4-P Chart

1. Preview	2. Predict	3. Prior Knowledge	4. Purpose

KWL Chart

K What I KNOW	W What I WANT to Know	L What I LEARNED

SQ3R Chart

S Survey	Q Question	R Read	R Recall	R Reflect

Semantic Web

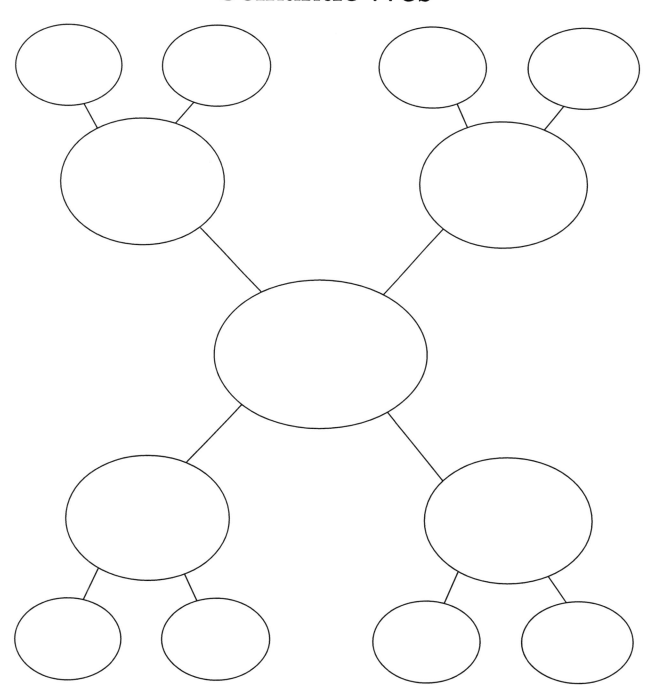

Outline

I. _____

    A. _____

        1. _____

        2. _____

        3. _____

    B. _____

        1. _____

        2. _____

        3. _____

II. _____

    A. _____

    B. _____

    C. _____

Structured Notes (some options)

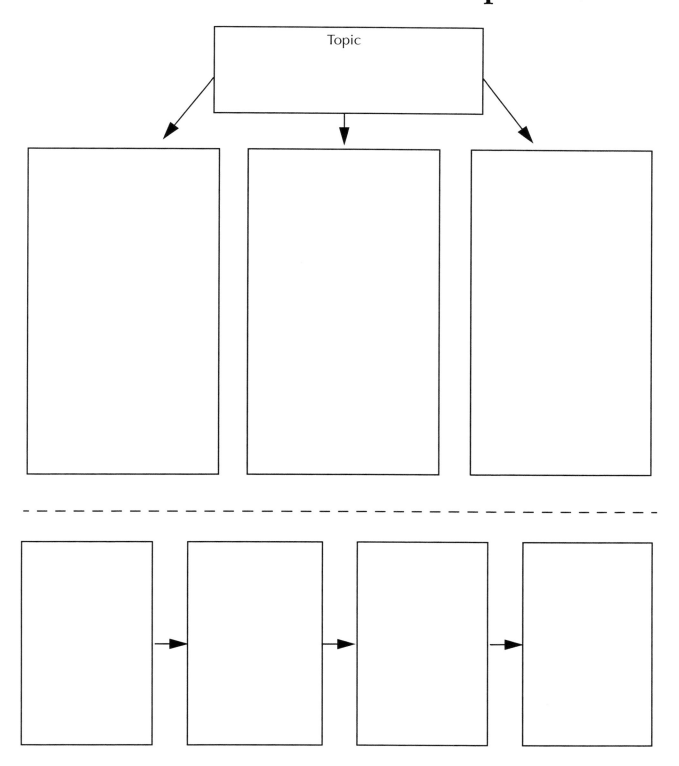

Teacher's Guide and Answer Key

Part 1: Building Vocabulary

Since so much of reading depends on understanding vocabulary, the first three lessons have been devoted to teaching strategies that help students decipher unfamiliar words. These strategies apply to any reading situation. Some of the vocabulary words, prefixes, suffixes, and roots are found particularly in mathematical material.

Lesson 1: Using Context Clues

1. (c) Steps 4, 5, 6, 7, and 8 refer to intersections as points.
2. (b) The second and third paragraphs talk about repeated use.
3. (d) The drawing makes it clear.
4. (a) The sentence refers to the 11-step procedure just given.
5. (b) The final drawing shows what the word means.

Lesson 2: Prefixes and Suffixes

1. (a) *Inter-* means "between."
2. (a) *Tensive* is *tense + ive*.
3. (c) *Compressive* is *compress + ive*.
4. (b) *Crystalline* is *crystal + ine*, which means "having the characteristics of."
5. (d) *Cohesive* is *co + hesive*. *Co-* means together. The rest of the word is similar to *adhesive*.

Lesson 3: Word Roots

1. multiplicative, commutative, associative, distributive
2. adjective
3. inverse: into + turn
   opposite: against + place
   multiply: many + fold
   product: forward + bring
   commutative: together + change
   divide: away + see
   associative: companionship + related to
   subtract: under + pull

Part 2: Prereading

This section focuses on prereading, the first of the three reading stages (prereading, reading, postreading). Students may already be familiar with the writing process, but the idea of reading as a process may be new. Seeing reading as a process rather than as a one-time task encourages students to invest the thought and time needed to make their reading worthwhile. They will eventually save time by using the strategies, while getting more from their reading.

Lesson 4: Previewing

In this lesson, students learn about and practice previewing a reading selection. Part of the previewing column of the 4-P chart has been filled in as a model. Other information that students record will vary, but there should not be too much specific information; at this stage, students should pay attention to heads, subheads, graphics, and key words found at the beginning and end of paragraphs. Additional information students might gather during this stage includes the following:

How to draw a right triangle—using circle; numbers 3, 4, 5 important in drawing triangles; drawings showing area; answered question about triangles.

Lesson 5: Predicting

In this lesson, students make predictions about the reading they will do, based on their previewing. Thinking ahead, asking questions about the text, and asking oneself questions all prepare the reader to take in and retain new knowledge by making reading an active process.

The questions and ideas that students write in the predict column of the 4-P chart will vary but may include some of the following:

The triangle drawing probably shows how to use that rope to make a triangle; importance of 3, 4, 5 will be explained; area is important for triangles—why?; Pythagoras asked a question about triangles—there will probably be a proof or an experiment showing how he figured it out.

Lesson 6: Prior Knowledge

Activating prior knowledge is an important step in the reading process. Students not only prepare themselves to make lasting connections between new and familiar material, which makes retaining the new information easier; they also gain confidence as readers and thinkers when they realize that they do have some knowledge of a topic. They start out with some sense of ownership of the material and increase that sense when they add to their knowledge.

Students' prior knowledge will vary. If they have trouble writing anything as prior knowledge, encourage them to free-associate on the following words: triangle, geometry, measuring, angle. Some students may have heard of Pythagoras, or know something about theorems.

Lesson 7: Purpose

This lesson discusses the reader's purpose. Of course, one purpose for reading is simply that it was assigned. Encourage students to go beyond the most obvious to consider what they hope to gain—what knowledge they hope to gather—from a particular reading selection.

This may be an appropriate time to discuss author's purpose and author's bias in writing. Although mathematics may seem to offer straightforward information without gray areas, it is useful to remind students that how something is presented can influence the person reading it. We have all heard that statistics can be twisted to show whatever someone wants to show, for example. By emphasizing one body of information and downplaying another, a writer could skew information.

Students expectations and opinions will vary.

Part 3: Reading Strategies

Lesson 8: Introduction to Reading Strategies

In this section of the book, five useful reading strategies are taught: KWL, SQ3R, semantic web, outline, and structured notes. These strategies use graphic organizers to help students organize their thinking, extract information from the reading, and retain meaning. Recording information and ideas on the graphic organizers strengthens both reading and writing skills. Visual representations of reading selections support visual learners and give students excellent study aids that they create themselves.

Lesson 9: KWL

The first reading strategy is the three-column KWL chart, which stands for What I Know, What I Want to Know, and What I Learned. The first two columns involve the prereading strategies of activating prior knowledge and defining a purpose for reading. Encouraging students to preview any text gives them more information to add to these columns; they can easily add what they know about the text from previewing and what they expect to know from predicting about the text.

1. (b)
2. (e)
3. (a)
4. (d)
5. Answers will vary. The following is a sample answer:

People might not like it if the sun did not rise and set at the times of day when they expect it to. In the time zones we have, on the equinoxes, the sun rises at about 6 A.M. and sets at about 6 P.M. If the entire continental United States were on Central Time, for example, the sun would rise at about 4 A.M. in California and at about 7 A.M. in the east. In the summer, of course, it would rise much earlier in every time zone.

Lesson 10: SQ3R

1. (a)
2. (c)
3. (d)
4. (a)
5. Answers will vary. The following is a sample answer:

The bees don't know that the hexagon is the winner. They only know how to make honeycombs with hexagonal cells. Bees that make hexagonal honeycombs are the winners in a long-running competition among bees and other organisms for survival. The winners get to pass their genes, which include DNA-encoded instructions for making hexagonal honeycombs, on to the next generation. If there were ever bees that made triangular or square honeycombs—and we don't know that there ever were—they must have died out. They probably spent too much time making wax and didn't have enough time left to make honey.

Lesson 11: Semantic Web

1. (b)
2. (c)
3. (b)
4. (c)
5. Answers will vary. The following is a sample answer:

   The set of all subatomic particles in the universe is probably not infinite in the strict sense of the definition given in the text. We certainly could not ever prove that it is. On the other hand, it is infinite in the sense that it is unimaginably large and impossible to count.

Lesson 12: Outline

1. (b)
2. (a)
3. (b)
4. (d)
5. Answers will vary. The following is a sample answer:

   You would never know what time it was anywhere on Earth that was not at exactly the same longitude as you unless you calculated their local mean time based on the difference in longitude between your location and theirs. To arrange to meet someone at a certain time and place, you would have to calculate not only travel time but also the time difference between places. The difference would almost always be a fraction of an hour, not a whole hour. Bus, train, plane, and television program schedules would be difficult to understand and use.

Lesson 13: Structured Notes

1. (c)
2. (c)
3. (c)
4. (c)
5. Answers will vary. The following is a sample answer:

   Scientific notation is a concise way to write and calculate with very large and very small numbers, and it is a good way to indicate significant figures. However, scientific notation is not as easy to read as ordinary notation. So when dealing with measurements in the normal range and when significant figures are not important or are understood, there's no point in using scientific notation.

Part 4: Postreading

Often, the only postreading activities students engage in involve assessment. Developing the habit of summarizing or paraphrasing what they have read helps students prepare for an assessment, as well as crystallizing new knowledge. Both the KWL and SQ3R reading strategies incorporate postreading. When students use the other strategies, encourage them to summarize or paraphrase their reading, using their graphic organizers for reference.

Lesson 14: Summarizing and Paraphrasing

There are, of course, many ways to summarize or paraphrase a text accurately. The following are samples:

Summary: The money a bank pays you for the use of your money is called interest. There are two kinds of interest: simple and compound. To find simple interest, multiply the principal, the amount of money invested, by the interest rate ($p \times r$). For example, the simple interest on $5000 invested for a year at 8% per year is $5000 \times .08 = $400. Compound interest means that the interest is added to the principal each year (or possibly more often). The 8% interest for the second year is paid on a principal of $5400, not $5000. $5400 \times .08 = $432. The $432 is added to the principal in the third year. Compounding the interest makes a big difference over a long period of time.

Paraphrase: If you have money to invest, the difference between simple and compound interest makes a big difference to you. Suppose you put $10,000 in a savings account at 4% a year. At the end of a year, the bank will give you $10,400 back if you want it. That's simple interest. Now suppose you leave your $10,400 in the bank. At the end of the second year, what will you have? $10,800? No. You'll have $10,816. At the end of the third year, you'll have $11,248.64. At the end of the fourth year, you'll have $11,698.58. That's compound interest. The interest you earn in one year is added to the next year's principal. The longer you leave it in the bank, the faster it grows.

Part 5: Reading in Mathematics

Lesson 15: Common Patterns and Features in Mathematics Writing

This section introduces students to some of the common patterns and features they are likely to encounter in reading for math. Recognizing patterns can help students choose which reading strategy to use.

Lesson 16: Main Idea and Details

Make sure that students understand that the main idea of a paragraph is not always clear-cut. Sometimes reasonable people can reasonably differ about it. They should also notice that a sentence that carries a main idea, like the first sentence of the second paragraph, could have details in it. Also, they should understand that the main idea of a paragraph or text is not always a statement; as in the fourth paragraph of the reading selection, it can be a question.

1. "But a solar year . . ."
2. "To make a calendar . . ."
3. "Three quarters of an hour . . ."
4. "Does the Gregorian calendar . . ."
5. "That's a problem for people who make calendars."

Lesson 17: Visual Texts

This lesson emphasizes the importance of visual literacy. Not only do students need to know how to read tables and graphs to learn the information contained therein, they also have to think critically about the visuals, about the author's purpose and bias, the source of the visual, and so on.

Answers will vary to the open-ended question in this lesson: Did any of the information in the text or visuals surprise you? So far as we know, the statements in the text are fair representations of the data in the table and graphs. Whether they are surprising is, of course, a subjective matter. Many people believe, perhaps because of media concentration on events like the shootings at Columbine High School, that serious youth crime is increasing. You might want to ask your students what they believe about that *before* they read this text.

Lesson 18: Concept and Definition

Answers will vary. Here are some more examples of converse relations:

higher—lower	teacher—student
older—younger	author—work
parent—child	multiple—factor
employer—employee	square—square root

Lesson 19: Concept on Concept

1. The order of concepts is *number sequence, terms, arithmetic sequence, common difference.*
2. Answers may vary.
3. Answers may vary.
4. Answers will vary.

Lesson 20: Steps in a Process

The following is a sample flowchart:

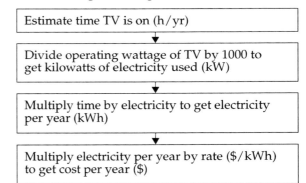

T = the time the TV set is on in hours per year

W = the operating wattage of the TV set

R = the rate that the electric company charges in dollars per kilowatt-hour

C = the cost of operating the TV set for a year in dollars

$$C = \frac{TWR}{1000}$$

Part 6: Practice Readings

This section of the book provides longer readings for students to practice their prereading,

reading, and postreading strategies. You may assign a particular graphic organizer or allow students to choose. A quiz follows each reading; the answers follow.

Practice Reading A: Supply and Demand

Any of the reading strategies works well for this reading.

1. (c)
2. (c)
3. (c)
4. (c)
5. Answers will vary. The following is a sample answer:

    If the wheat crop were largely wiped out by pests, there would only be a small amount of wheat available to be sold and the price would be very high.

Practice Reading B: The Three Suspects

Perhaps the most useful reading strategy for this reading is structured notes showing the chief's thought process—some kind of flowchart.

1. (a)
2. (c)
3. (a)
4. (c)
5. Answers will vary. The following is a sample answer:

    The premises that one person always tells the truth and another person never tells the truth are not realistic. To always tell the truth or never tell the truth requires complete knowledge of the truth, which is not granted to ordinary mortals. However, anecdotes of this kind, although unrealistic, provide interesting starting points for exercises in logical reasoning. What *is* realistic is that many serious discussions in science, business, government, and other fields begin with a set of premises that are accepted as true by all participants and proceed by logical reasoning to conclusions.

Practice Reading C: Sampling Red Blood Cells

Any of the reading strategies work for this reading. Perhaps structured notes are the most useful, depending on students' ability to see and make use of the reading's structure.

1. (a)
2. (a)
3. (d)
4. (c)
5. This is not a mistake. It doesn't matter. The difference between 199 and 200 is well within the sampling error.

Practice Reading D: Archimedes

KWL, SQR3, or outline may be good choices of strategies for this reading.

1. (a)
2. (a)
3. (b)
4. (c)
5. Answers will vary.